URANIUM

AND OTHER RADIOACTIVE ELEMENTS

U

Atlantic Europe Publishing

How to use this book

This book has been carefully developed to help you understand the chemistry of the elements. In it you will find a systematic and comprehensive coverage of the basic qualities of each element. Each two-page entry contains information at various levels of technical content and language, along with definitions of useful technical terms, as shown in the thumbnail diagram to the right. There is a comprehensive glossary of technical terms at the back of the book, along with an extensive index, key facts, an explanation of the Periodic Table, and a description of how to interpret chemical equations.

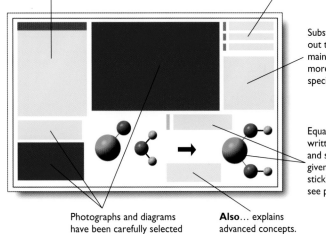

The main text follows the sequence of information in the book and summarises the concepts presented on the two pages.

Technical definitions.

Substatements flesh out the ideas in the main text with more fact and specific explanation.

Equations are written as symbols and sometimes given as "ball-and-stick" diagrams – see page 48.

Photographs and diagrams have been carefully selected and annotated for clarity.

Also… explains advanced concepts.

· ·

An Atlantic Europe Publishing Book

Author
Brian Knapp, BSc, PhD
Project consultant
Keith B. Walshaw, MA, BSc, DPhil
    (Head of Chemistry, Leighton Park School)
Industrial consultant
Jack Brettle, BSc, PhD (Chief Research Scientist, Pilkington plc)
Art Director
Duncan McCrae, BSc
Editor
Elizabeth Walker, BA
Special photography
Ian Gledhill
Illustrations
David Woodroffe and David Hardy
Designed and produced by
EARTHSCAPE EDITIONS
Print consultants
Landmark Production Consultants Ltd
Reproduced by
Leo Reprographics
Printed and bound by
Paramount Printing Company Ltd

First published in 1996 by
Atlantic Europe Publishing Company Limited, Greys Court Farm,
Greys Court, Henley-on-Thames, Oxon, RG9 4PG, UK.

Suggested cataloguing location
Knapp, Brian
    Uranium and other radioactive elements
    ISBN 1 869860 69 1
    – Elements series
540

Acknowledgements
The publishers would like to thank the following for their kind help and advice: Rolls-Royce plc and Jonathan Frankel of J.M. Frankel and Associates.

Picture credits
All photographs from the **Earthscape Editions** except the following: (c=centre t=top b=bottom l=left r=right)
Hulton Deutsch Collection 39br; coutesy of **Rolls-Royce plc** 41tl; **Science Picture Library** 21 (Will & Deni McIntyre), 24/25b (J. C. Revy); **UKAEA Technology** FRONT COVER, BACK COVER (br, cr, tr & cl), 4/5, 9t, 12t, 12b, 13tl, 18/19b, 19tl, 22/23, 26b, 26/27, 32/33t, 33cl, 36/37, 37t, 41br, 42/43b, 43c, 44b, 44/45t, 45b and **ZEFA** BACK COVER (bl) and TITLE PAGE (Photri), 24/25t (Jim Brandenburg), 38/39 (Photri).

Front cover: A bundle of tubes which will be filled with uranium metal or uranium dioxide fuel (fuel pins) makes up this fuel element from an advanced gas-cooled reactor. When filled with fuel, these elements are then placed into the reactor core. One fuel element produces as much energy as about three thousand tonnes of coal.
Title page: A nuclear bomb test.

Contents

Introduction

An element is a substance that cannot be broken down into a simpler substance by any known means. Each of the 92 naturally occurring elements is therefore one of the fundamental materials from which everything in the Universe is made. This book is about uranium and other radioactive elements.

Radioactive elements

In chemical reactions, the atoms that make up the elements do not change. However, in some circumstances the atoms can be made to change (such as when two hydrogen atoms are made to fuse together to produce a helium atom) and some elements, such as uranium, continually change. Elements that continually change are called radioactive elements.

All atomic change is characterised by a release of energy, which may be in the form of heat alone, heat and light together, or, as in the case of the radioactive elements, heat, light and the release of "radiation" (rays or particles of matter).

All the elements with an atomic number of 88 and above in the Periodic Table (see page 46) are radioactive. In addition, some common elements (such as hydrogen, oxygen and carbon) have radioactive forms called isotopes.

The changes in radioactive elements take place in the core, or nucleus, of the element's atoms. This is why scientists call them nuclear reactions and why we use such terms as nuclear energy and nuclear bombs.

Nuclear reactions are often far more powerful (they release much more energy) than chemical reactions. This is why they have been used in bombs as well as in power stations. However, when properly handled, they are not any more dangerous than chemical reactions.

Radioactivity – energy given out by a radioactive substance – is a result of changes to the nucleus of an atom. The atoms of the radioactive elements send out, or radiate, particles and waves of energy from their core (nucleus) without any form of chemical reaction.

Radioactivity is as old as the Universe, yet it was only discovered quite recently. Radium, the first radioactive element to be discovered, was only noticed in 1896 when French scientist Antoine Henri Becquerel accidentally left an unexposed photographic plate in a drawer next to a piece of ore which contained radium. When he developed the plate he found that it had been "fogged" by the radioactivity from the radium. The word radioactivity was suggested in the early years of this century by Pierre and Marie Curie, who did much of the pioneering work on radioactive elements.

Radioactive elements are not stable like other elements. Because they continuously give out radiation, they are also always changing. Although radium and uranium are probably the best known of the radioactive elements, there are many others with very useful properties, as we shall see.

◀ Testing the construction of the fuel rod assembly before inserting the radioactive material.

Inside the elements

Chemistry is concerned with the exchange or sharing of electrons in the outside regions of an atom, rather than with what happens in the core of an atom, in the nucleus. You could compare this to a traffic planner who is concerned only with the way vehicles move, not with how their engines work.

However, when working with radioactive elements, scientists, and especially physicists, need to understand the nature of the particles that make up the atoms, much as a vehicle-maker needs to know how engines work.

Radioactive elements vary greatly. All are quite rare; most are metals. But radioactive isotopes cannot be used in the same way as other metals – to make new materials. Indeed, if used this way they could make rather dangerous materials. Rather, radioactive elements are used for their radioactive properties alone.

Unstable nucleus —

▶ A radioactive atom, showing a particle being emitted from the nucleus. The other diagrams in this book are representations of the nucleus and not the whole atom.

Particle emitted by the nucleus

Electrons orbit the nucleus

What makes an element radioactive?

Inside an atom there are three kinds of particle: protons, neutrons and electrons. The nucleus, the tiny core of the atom, contains protons (positively charged particles) and neutrons (so called because they are neutral and have no charge). The region beyond the nucleus contains (negatively charged) electrons that balance out the charge of the protons. The electrons are usually thought of as orbiting the tiny nucleus, like planets orbiting the Sun.

As you would expect, properties of an atom give rise to some of the most important properties of an element. For example, the atomic number of an element is equal to the total number of protons in an atom (the atomic number of each element is given on pages 46–47). The total number of protons and neutrons gives the atomic weight (mass). There are roughly as many protons as neutrons, which is why the atomic weight is about (but not exactly) twice the atomic number. In some cases several versions of the same element occur, identical in all their chemical properties and varying only in the number of neutrons they contain. Each of these variations is called an isotope (meaning "the same but different").

Because like charges repel each other, there is always a force trying to push the protons apart. Provided there are not too many protons in the nucleus, other forces can hold the protons together. But if the ratio of protons to neutrons is not within certain limits, protons may not be held firmly together, and they form an unstable nucleus. This is what makes isotopes of some elements radioactive.

For example, carbon, the element found in all living things has the chemical symbol C. The normal form (isotope) has an atomic weight of 12 and is written ^{12}C or carbon-12, but the radioactive version (the radioactive isotope) has two extra neutrons, so its symbol is ^{14}C (carbon-14). As we shall see, the radioactive form behaves chemically just like the non-radioactive form, although one will never change into the other.

atom: the smallest particle of an element.

electron: a tiny, negatively charged particle that is part of an atom. The flow of electrons through a solid material such as a wire produces an electric current.

isotope: atoms that have the same number of protons in their nucleus, but which have different masses; for example, carbon-12 and carbon-14.

neutron: a particle inside the nucleus of an atom that is neutral and has no charge.

proton: a positively charged particle in the nucleus of an atom that balances out the charge of the surrounding electrons.

Also...
The origin of the elements

Why do we have so many different elements and where did they form? This is a question that a nuclear scientist is best able to answer, because the answer lies in the core, or nucleus, of atoms.

At the beginning of time (the instant of the creation of the known Universe, called the Big Bang) the only element in existence was hydrogen. All the other elements are, in some way or another, "daughters" of hydrogen. So, for example, hydrogen atoms are forced together (fused) to make helium and so on down a long line. In this line carbon, for example, is transformed into oxygen. Together the elements created by fusion make the stuff of life.

Nuclear reactions are the most fundamental of all reactions, creating the elements themselves. They are going on today, just as they have been happening since the Big Bang. Most nuclear activity occurs in the stars, although a small amount is happening in the rocks of the Earth. But even nuclear reactions on a small scale are noticeable because of the outpouring of energy – radiation – that accompanies any nuclear change.

Types of radioactivity

There are a number of ways in which a nucleus can changes. Each change shown here produces a different type of radiation.

Alpha particles

The nucleus breaks down. A stable combination of two protons and two neutrons (known as an alpha particle) is ejected from the nucleus as it decays.

As it happens, an alpha particle is also the nucleus of the atom of helium. If it captures two electrons, it can become a neutral helium atom. It does this by crashing into nearby atoms. All alpha particles readily transform into helium atoms.

The remaining atom has less mass and less charge than before, so it becomes a new element with a lower atomic number. This process typically occurs in heavy elements like uranium. In this way unstable radioactive uranium eventually changes (decays) to non-radioactive, stable lead.

Beta particles

The nucleus breaks down and ejects an electron (which is called a beta particle). What remains is a new element with a higher atomic number.

This commonly happens in light elements. Thus, for example, tritium (an isotope of hydrogen) breaks down into helium, carbon changes to nitrogen, and nitrogen changes to oxygen.

Gamma rays

The nucleus breaks down and rearranges itself into a tighter cluster, sending out a wave of energy. The wave of energy is called gamma radiation. It is the same type of radiation as X-rays. Gamma rays carry enough energy to damage cells. This is how they kill living matter. It is for this reason that radioactive sources have to be shielded behind some absorbing material such as lead.

Neutrons

The nucleus breaks down and emits neutrons. "Streams" of neutrons were first observed when light elements such as beryllium were placed next to radioactive substances that emitted alpha particles.

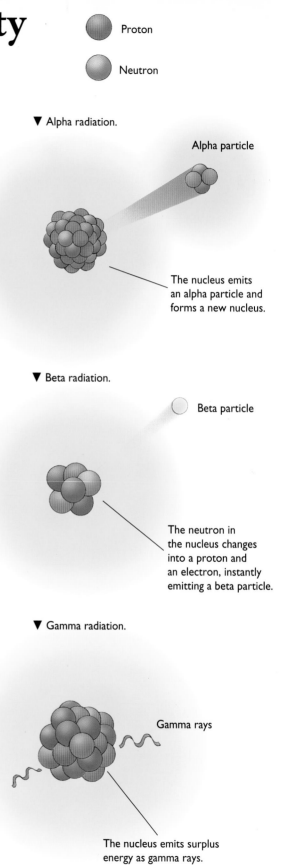

Proton

Neutron

▼ Alpha radiation.

Alpha particle

The nucleus emits an alpha particle and forms a new nucleus.

▼ Beta radiation.

Beta particle

The neutron in the nucleus changes into a proton and an electron, instantly emitting a beta particle.

▼ Gamma radiation.

Gamma rays

The nucleus emits surplus energy as gamma rays.

▲ In 1917, scientist Ernest Rutherford was the first person to change one element into another. He transformed nitrogen into oxygen. In this picture, Lord Rutherford (centre) is accompanied by Dr E. S. Walton (left) and Dr J. D. Cockroft (right).

▼ Different forms of radiation have different powers of penetration. Alpha particles are the least penetrating, and reach only a few centimetres from the radioactive source. This is why plutonium and uranium can be handled safely with rubber gloves. Beta particles will travel a few metres and penetrate bodies, but are stopped by metals. Gamma rays (emitted by, for example, cobalt-60) will travel for several kilometres from the radioactive source and they will penetrate through most metals. They are only stopped by lead and thick concrete. Neutrons can be absorbed by dense polythene, water, concrete (which contains water) and other compounds that contain hydrogen.

alpha particle: a stable combination of two protons and two neutrons, which is ejected from the nucleus of a radioactive atom as it decays. An alpha particle is also the nucleus of the atom of helium. If it captures two electrons it can become a neutral helium atom.

beta particle: a form of radiation in which electrons are emitted from an atom as the nucleus breaks down.

electron: a tiny, negatively charged particle that is part of an atom. The flow of electrons through a solid material such as a wire produces an electric current.

gamma rays: waves of radiation produced as the nucleus of a radioactive element rearranges itself into a tighter cluster of protons and neutrons. Gamma rays carry enough energy to damage living cells.

nucleus: the core of an atom, a tiny region where protons and neutrons are held together.

neutron: a particle inside the nucleus of an atom that is neutral and has no charge.

proton: a positively charged particle in the nucleus of an atom that balances out the charge of the surrounding electrons.

radioactive decay: a change in a radioactive element due to loss of mass through radiation. For example uranium decays (changes) to lead.

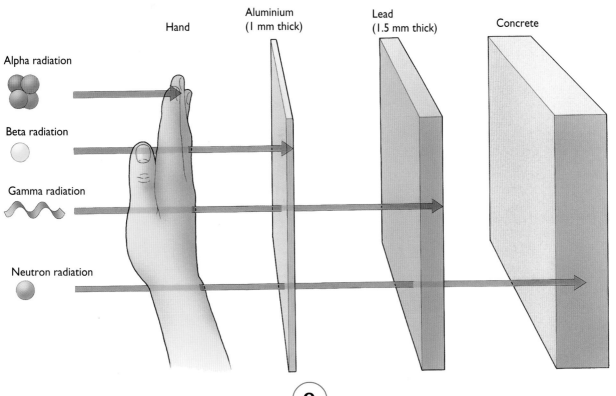

Hand — Aluminium (1 mm thick) — Lead (1.5 mm thick) — Concrete

Alpha radiation

Beta radiation

Gamma radiation

Neutron radiation

Changing one element into another

The change from one element to another is called transmutation. Remember that atoms are made up of different combinations of protons, neutrons and electrons, so the break up of a neutron leaves behind an atom with a different combination of neutrons, protons and electrons.

This new combination is a new element. So when uranium loses nuclear particles, a whole chain of events is set in motion until finally lead is created. Different forms of radiation are produced during this process.

The chain of events tends to happen very slowly, sometimes only over many millions of years. As we shall see, this slow rate of change gives rise to one of the main problems of dealing with radioactive materials.

In nature, new radioactive materials are produced all the time. This is because particles from space (known as cosmic rays) are continually bombarding the air and the planet's surface (including our bodies). The particles they contain have enough energy to change nitrogen-14 in the atmosphere to carbon-14, which is absorbed in living tissue in the same way as carbon-12. For this reason, carbon-14 can be used as a radioactive "clock" (see page 18).

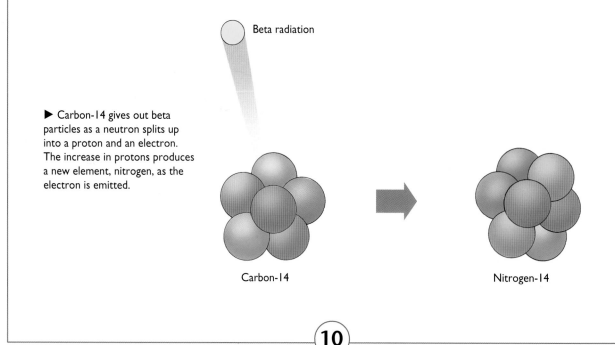

Beta radiation

► Carbon-14 gives out beta particles as a neutron splits up into a proton and an electron. The increase in protons produces a new element, nitrogen, as the electron is emitted.

Carbon-14

Nitrogen-14

Neutron

▼ Radioactive potassium changes to argon gas.

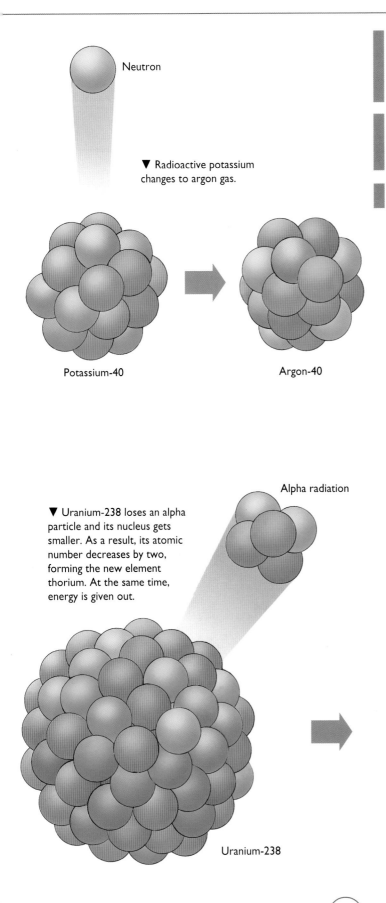

Potassium-40

Argon-40

Also...
The alchemist's dream

In the Middle Ages, alchemists worked to change lead (a common and cheap metal) into gold (a rare and expensive metal). They called the process they were seeking transmutation.

But no matter how hard they tried, they could not achieve this by chemical means. In fact, no chemical or mechanical process can change the nucleus of an atom. It can be hit with a hammer or boiled in acid without change. This is because in chemical reactions, there is not enough energy to change the nucleus of the elements themselves, only the way they are combined.

In a curious twist of fate, the alchemists little knew that nature had achieved transmutation since the beginning of time. The change from one element to another is a "simple" matter of changing the number of protons in the core of an atom. In fact, radioactive decay has taken place since the earliest moments of the creation of the Universe, changing one element into another, but not lead into gold!

Alpha radiation

▼ Uranium-238 loses an alpha particle and its nucleus gets smaller. As a result, its atomic number decreases by two, forming the new element thorium. At the same time, energy is given out.

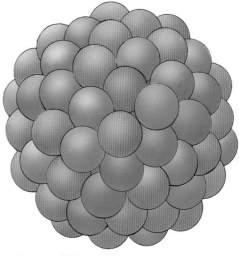

Uranium-238

Thorium-234

Measuring radioactivity

When Henri Becquerel first detected radioactivity, it was because it had caused a photographic plate to become fogged. It was quite natural, therefore, that Becquerel should use photographic film as a radiation detector. By seeing how much a film had become fogged, he was able to get a rough guide to the intensity of the radiation.

This technique is still used to measure the amount of radioactivity received by hospital or laboratory workers. The small badge that such people wear contains a film that is developed on a regular basis.

Geiger counter

One of the early workers in the field of radioactivity was Hans Geiger. He developed a sensitive radiation detector using a gas-filled tube.

The Geiger–Müller tube (Geiger counter) is a metal tube fitted with a thin metal wire at its centre. The tube is filled with gas (usually a mixture of neon and bromine gases) at very low pressure. A large voltage is applied between the central wire and the outer casing.

Normally nothing happens. But in the presence of radioactivity, radiation causes some of the gas atoms to break up (they ionise). This allows the gas in the tube to conduct electricity. The flow of electrical current can be detected on an electrical meter. In many Geiger counters the flow of electricity goes into a special circuit that makes clicks in a loudspeaker. The greater the intensity of the radiation, the more often current flows and the more clicks are heard each second.

Each of the types of radiation has a different power of penetration, so cylinders can be designed with "windows" – different thicknesses of body wall – in order to measure each type of radiation.

▲ A film badge measures the dose and type of radiation its wearer receives. Like the Geiger counter, badges have different "windows" (thicknesses of shielding over different parts of the film) for this purpose.

▼ A modern Geiger counter.

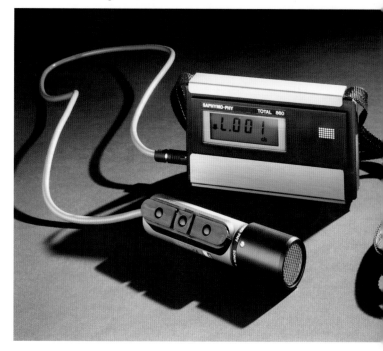

becquerel: a unit of radiation equal to one nuclear disintegration per second.

curie: a unit of radiation. The amount of radiation emitted by 1 g of radium each second. (The curie is equal to 37 billion becquerels.)

ion: an atom, or group of atoms, that has gained or lost one or more electrons and so developed an electrical charge. Ions behave differently from electrically neutral atoms and molecules. They can move in an electric field, and they can also bind strongly to solvent molecules such as water. Positively charged ions are called cations; negatively charged ions are called anions. Ions carry electrical current through solutions.

◀ Using a Geiger counter to detect radioactive emissions on contaminated waste land.

Also...

The Geiger counter has been used to measure the amount of radiation emitted by one gram of radium each second. This value, 37 billion pulses of radiation per second, has been used to define a unit of radiation called the curie. Another unit is the becquerel, one pulse of radiation per second.

The outer metal tube acts as the cathode.

Radiation

The thin metal wire acts as an anode.

The metal tube is filled with neon and bromine gasses at low pressure.

▶ This diagram shows the components of a Geiger counter.

450 V supply, amplifier and counter (which measures the counts per minute)

Background radiation

Radioactivity is a fundamental part of the Universe, and therefore we should expect to find natural radiation on Earth. In fact we experience radiation from a variety of sources as a normal part of our lives. Together, radiation from these sources makes up "background radiation".

The levels of background radiation have been increased by about one-fifth because of human activities this century, although the vast majority of background radiation is still from natural sources. A few people are more at risk than others from background radiation, especially those living in houses built above rocks rich in radium and from which radon gas seeps to the surface.

Cosmic radiation

This is a term for all the forms of radiation that reach the Earth from other parts of the Universe, including the Sun and all other stars.

Victor Franz Hess, an Austrian physicist, was the first person to prove that radiation continually reached the Earth from space. He received the Nobel Prize in 1936 for this discovery.

Hess made measurements in balloons that showed that the amount of radiation increased with height. At the height an aircraft travels, the radiation is several times that on the Earth's surface. This means that frequent aeroplane flyers are slightly more exposed to radiation than those who never fly.

Most cosmic rays are protons, but there are also alpha and beta particles. Cosmic radiation can be of the order of 1 particle/sq cm/second.

Radioactive carbon dioxide

Cosmic rays bombard all the gases in the upper atmosphere. One effect is to change nitrogen-14 to carbon-14, the radioactive isotope of carbon. When carbon-14 and oxygen combine in the upper air, they produce radioactive carbon dioxide. Some of this is absorbed by plants and can be used to determine their age (see page 19).

▼ This pie chart shows the main sources of radiation.

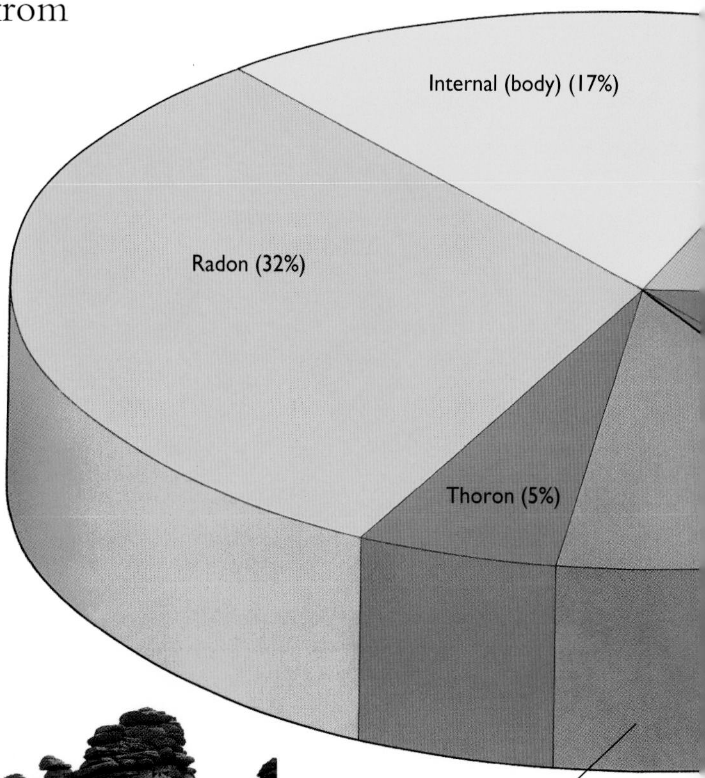

Natural background radiation from the ground (about 37%)

Internal (body) (17%)

Radon (32%)

Thoron (5%)

Medical (11.5%)

▶ Radon gas is most concentrated over igneous rocks such as granite.

Radiation from man-made sources is dominated by X-rays and fallout from nuclear reactors (about 13%).

radioactive decay: a change in a radioactive element due to loss of mass through radiation. For example uranium decays (changes) to lead.

X-rays: a form of very short wave radiation.

◄ Medical X-rays are an important contributor to man-made radiation.

Radiation from the ground

Radioactive elements in rocks are found in groundwater and soils. They produce about half of the natural background radiation.

Radium, a natural element found in many rocks, especially those associated with granite and other volcanic materials, decays to release life-threatening radon gas.

Man-made radiation

Just under one-fifth of the radiation we receive is from many everyday sources, such as having an X-ray in a hospital. Other factors contributing to low level background radiation include some radioactive gases and liquids released by nuclear power stations. Very occasionally there is an increase in this background radiation due to nuclear weapons testing in the atmosphere or an accident in a nuclear power station.

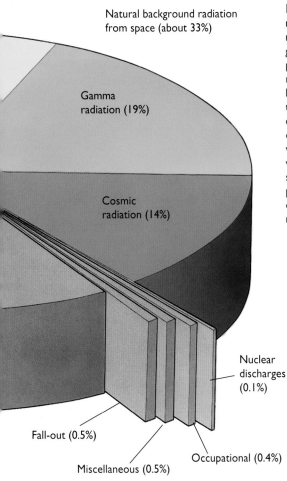

Natural background radiation from space (about 33%)

Gamma radiation (19%)

Cosmic radiation (14%)

Nuclear discharges (0.1%)

Fall-out (0.5%)

Miscellaneous (0.5%)

Occupational (0.4%)

► Smoke detectors use a minute amount of radioactive material separated by a small air gap from a detector. When smoke particles enter the air gap, they reduce the count rate registered by the detector, and the alarm is triggered. This provides a cheap and effective means of helping to prevent deaths by fire. Such detectors emit very low energy beta radiation, which is easily absorbed by the surrounding air and so does not present a health hazard or contribute significantly to the normal background radiation.

Sensing chamber containing radioactive source

Our radioactive bodies

Our own bodies have many atoms within them, some of which are naturally radioactive. The most common of these is radioactive potassium, making up about one in ten thousand atoms of potassium. As these atoms decay, they release about a tenth of a microcurie, meaning that 3700 nuclear disintegrations take place inside your body every second. Thus about 17% of the background radiation we receive occurs inside ourselves!

Half-life

Radioactive materials are widely scattered in all the rocks of the Earth. Pick up any volcanic rock from the ground and it is likely to contain atoms of radioactive elements.

Radioactive elements send out radiation all the time, although gradually the level of radiation decreases. However, whatever the level of radiation, all radioactive materials lose half of their remaining surplus energy at a fixed rate. Thus, the time taken for a piece of an element to lose half of its remaining energy is always the same, whether the radioactivity is strong or weak.

The rate of change is described by the term half–life, meaning the time it takes for the radiation of an element to decrease by half.

Decay: for ever

Any very radioactive material will emit with the greatest intensity (and so be at its most dangerous) for a relatively short time after it has formed because, as the chart shows, it decays by fission fastest soon after its formation. This means that the useful energy from a radioactive element lasts for a relatively short time. Thus fuel rods in nuclear reactors have to be changed from time to time (see page 34).

The element may lose much of its energy at an early stage in its decay, but it never becomes fully stable, or finishes decaying. Thus radioactive elements remain slightly radioactive for ever. This is why there is a problem in dealing with radioactive materials that contain long-lived isotopes; the long half-life means that the material takes a long time for the level of radioactivity to fall to a trivial level. Some sources include warheads from nuclear devices, spent fuel rods from nuclear power stations and spent sources from medical equipment.

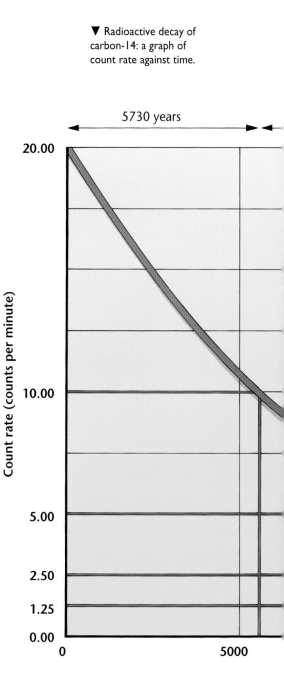

▼ Radioactive decay of carbon-14: a graph of count rate against time.

Half-lives

The half-lives of radioactive elements show dramatic differences. Most artificial radioisotopes have very short half-lives. Those radioisotopes that have naturally short half-lives have decayed away earlier in the history of the Earth. As a result, natural radioisotopes are dominated by those species with long half-lives. The only exceptions to this are radioactive isotopes like carbon-14, which are regenerated by cosmic rays in the atmosphere.

fuel rods: rods of uranium or other radioactive material used as a fuel in nuclear power stations.

half-life: the time it takes for the radiation coming from a sample of a radioactive element to decrease by half.

radioisotope: a shortened version of the phrase radioactive isotope.

Element	Half-life
Carbon-14	5730 years
Cobalt-60	5.3 years
Hydrogen-3 (Tritium)	12.3 years
Polonium-210	138 days
Radium-226	1600 years
Uranium-238	4.5 billion years

▶ The half-life of radioisotopes can vary greatly.

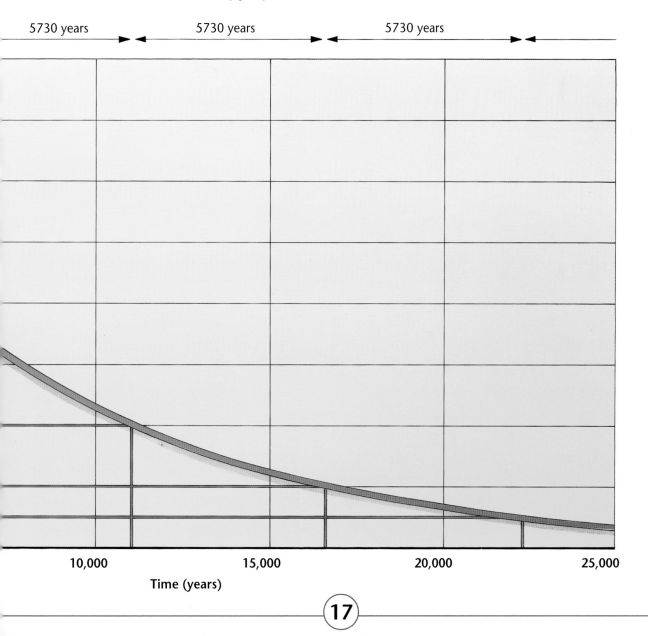

5730 years   5730 years   5730 years

Time (years)

Radioactive clocks

Every radioactive element decays by emitting radiation and particles until it changes into a stable element. Sometimes it passes through a series of transformations into other radioactive elements before finally becoming stable. Each radioactive element also has its own steady rate of decay.

To see how this happens, consider the following example. Suppose a piece of an element loses half its surplus particles and radiation in the first 10 years after it has formed. This is taken as the "tape measure" for radioactive decay, or half-life (see page 16). Half of what remains would be lost over the next ten years. At the end of 20 years there would be half of a half, or one-quarter of the starting amount. Ten years later there would be half of one-quarter (one-eighth) of the starting amount and so on. Because of this constant and predictable *rate* of change, it is possible to use it as an extremely accurate clock.

Several radioactive elements have very long half-lives (they decay very slowly) and they can be used for dating rocks. For example, radioactive potassium occurs in minerals such as hornblende and mica, and these minerals are found very widely in igneous rocks.

Rubidium is another radioactive element found in ancient volcanic rocks. The decay of rubidium to the stable element strontium has been used to date the world's oldest surviving rocks (at 3.8 billion years).

The Moon's rocks have also been dated by this method, providing us with a date for the origin of the Solar System of about 4.6 billion years ago.

▲ The minerals from igneous rocks such as granite and basalt contain radioisotopes. Their decay can be accurately measured to estimate a date for the rock formation. This sample is granite, which is widely found on continents. Basalt is the most common rock of the ocean floors. By using such samples, both land and sea rocks can be dated to help build a clearer picture of geological change.

▼ These archaeologically significant human remains were trapped in the peaty fibres of a marsh. By the use of carbon dating it was possible for scientists to ascertain the exact age of "Pete Bog".

basalt: an igneous rock with a low proportion of silica (usually below 55%). It has microscopically small crystals.

granite: an igneous rock with a high proportion of silica (usually over 65%). It has well-developed large crystals. The largest pink, grey or white crystals are feldspar.

half-life: the time it takes for the radiation coming from a sample of a radioactive element to decrease by half.

igneous rock: a rock that has solidified from molten rock, either volcanic lava on the Earth's surface or magma deep underground. In either case the rock develops a network of interlocking crystals.

◄ The round table was thought to have belonged to King Arthur of England and therefore date from the 6th century. Carbon dating revealed that it was, in fact, made in the 13th century.

Carbon-14

Carbon-14 is a radioactive isotope found in living things. It is produced high in the atmosphere where nitrogen-14 is bombarded by cosmic rays.

All forms of carbon dioxide are absorbed by plants and used to form tissues. As long as the plants, or animals that feed on them, are alive, there is a constant amount of radioactivity in the tissues, because old, decaying carbon-14 is replaced by further high activity carbon-14 from the air. As soon as the organism dies, the carbon is no longer absorbed and so the amount of radioactivity begins to decrease.

By measuring the radioactivity of carbon in a sample of organic material – timber, cloth, seeds, mummified bodies, etc. – its age can be accurately determined. Carbon-14 has a long half-life (about 5600 years). It can readily be used to date materials that are up to ten thousand years old – and a few even older samples have also been dated.

Radioactive tracers

Many radioactive materials are used to trace how chemical reactions have taken place, what happens to materials in nature and what happens inside the body. These are called radiotracers.

The important factor in the use of radiotracers is that both a radioactive isotope and its stable counterpart behave in much the same way in the environment. And because radiotracers can be detected at extremely low concentrations (sometimes down to one part in a trillion), it is possible to tag material with radiotracers even for such sensitive investigations as medical diagnosis.

A wide range of tracing isotopes is available for use. In farming research, for example, scientists label fertilisers with radioactive nitrogen, potassium or phosphorus to examine how plants take up nutrients from the soil.

In medical diagnosis, radioactive iodine is used by doctors to work out how fast iodine is taken up by the thyroid gland. Our bodies need some iodine, and when the rate is low, people may suffer from goitres and other problems.

Radioactive technetium, a synthetic element, is used by doctors to find out how various organs work. For example, doctors can inject it into the bloodstream to find the location of brain tumours.

Sodium, as salt, is naturally present in the blood supply. Radioactive sodium is used to trace the way the blood flows through the arteries and veins. Radioactive strontium (which behaves like calcium in the bones) is used to trace bone growth.

Detecting pollution

Radiotracers are widely used to monitor the behaviour of pollutants in the environment. A small amount of radiotracer added to the discharge of a factory, for example, presents no health hazard in itself, and the tracer can be used to examine what happens to the effluent.

Tracing holes in metal

Companies that make high precision metalwork such as aircraft engines need to check to make sure there are no tiny holes in the metal they use or the welds they have made. X-ray examinations are often used for this purpose. Because X-rays are absorbed by metal, they cannot be used to examine very thick metals. In these cases, it is sometimes possible to look at the way oil behaves in an engine by using neutron radiography techniques. In this method a source of neutrons is used. Neutrons are absorbed by any hydrogen-containing material, such as oil, so even tiny leaks will show up as a change in the absorption of neutrons. Thus, possible leaks can be traced.

▶ Radioactive xenon can be used as a tracer for medical use. In this picture the patient is inhaling a gas mixture which includes radioactive xenon. This will concentrate in the blood vessels of the brain. Detectors in the helmet which is placed over the patient's head trace the gamma rays emitted by the tracers, and a computer is used to interpret the data and create a graphic image of the circulation in the patient's brain, as can be seen on the computer monitor in the background.

Tritium

Tritium (hydrogen-3) is a natural radioisotope of hydrogen, with a half-life of about 12 years. It is produced as cosmic rays bombard hydrogen atoms inside water vapour molecules in the upper atmosphere.

Tritium is washed out of the atmosphere and then continues on in the water cycle (i.e. through plants, soil, rocks and rivers to the oceans).

Atmospheric nuclear tests greatly increase the amount of tritium in the air, which can also be detected in the water cycle. This has allowed scientists to follow the slow movement of underground waters.

Tritium is used to make nuclear weapons – hydrogen bombs.

radioisotope: a shortened version of the phrase radioactive isotope.

radiotracer: a radioactive isotope that is added to a stable, nonradioactive material in order to trace how it moves and its concentration.

Tracers detect leaky pipes

Radioactive tracers can be fed into water or gas pipes and a Geiger counter used to detect places where the amount of activity is high. This will be where the tracer is leaking into the surroundings and becoming concentrated.

By using this technique, long stretches of ground do not have to be dug up unnecessarily.

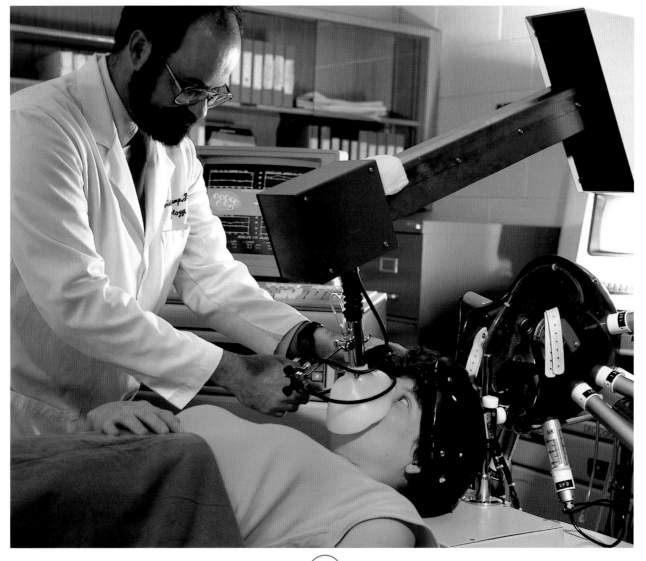

Radium

Radium (chemical symbol Ra) is a radioactive element with chemical properties similar to those of calcium and barium. Its name comes from the Latin *radius*, meaning "ray". Radium was the first radioactive element ever discovered, and the word radiation also comes from the same Latin source.

Radium is found combined with the more widely known radioactive element uranium. In 1896 Antoine Henri Becquerel discovered by accident that a compound of uranium caused photographic plates to become fogged just as though they had been exposed to light. From this observation it became clear that the ore was sending out invisible rays that had the same effect on the photographic plate as light rays. The difference was that these rays could penetrate materials such as paper and wood and thus fog a photographic plate even though it had been carefully wrapped up.

Marie and Pierre Curie refined many tonnes of uranium ore (pitchblende) before they found the element radium. (It is usual to find just one gram of radium in every seven tonnes of pitchblende!)

Radium is found alongside uranium because it is one of the elements formed by the transmutation of uranium as it decays (see page 10). However, far more uranium will already have become lead, so that only one part in three million of uranium has still to become radium (and this, in time, will also become lead).

Because radium was one of the first radioactive elements to be examined, the way it changes with time has been used as the measure of radiation. The units of radiation, the becquerel and the curie, are named after these early pioneers.

Most compounds of radium are colourless when first prepared, but they become coloured upon standing, because the intense radiation causes changes in the atoms of the compounds.

Radium is little used today.

▲ Marie and Pierre Curie at work in their laboratory in France, 1903.

Radon

Radon is a radioactive gas that occurs naturally in rocks and soils. Radon is emitted from the radioactive decay of radium and has a half-life of about 3 days. It is responsible for about half of the natural background radiation on Earth (see page 14).

Radon levels can be high above areas where there are granite rocks. The gas can build up inside homes as it seeps from the ground. In such areas special precautions have to be taken to ventilate houses. If this is not done, then the radioactivity, breathed as gas into the lungs over a number of years, can increase the risks of cancer.

Marie and Pierre Curie

The husband-and-wife team of Marie and Pierre Curie is most commonly associated with the discovery of radioactivity.

Pierre Curie was interested in the nature of crystals and how their properties could be used. He became the head of the laboratory in the Ecole de Physique et de Chimie Industrielle in Paris. After their marriage, Marie Curie studied the strange nature of radioactivity that had been discovered accidentally in 1896 by Henri Becquerel. During this work the Curies discovered a new radioactive element, polonium (named after Marie's country of origin, Poland), and then radium. By 1903 the Curies and Henri Becquerel had been honoured with the Nobel Prize for Physics for discovering radioactivity.

After the accidental death of her husband, Marie became a lecturer in the Sorbonne University in Paris and received a second Nobel Prize in 1911, on this occasion in chemistry for finding out the properties of radium.

Later, Marie Curie become interested in X-rays and saw their potential for use in medicine. Sadly, her death from leukaemia was the price she had to pay for working so long with radioactive materials without the knowledge that radiation could be harmful to people.

The work of the Curies, coupled with other research that explained the structure of atoms, led the way to realising that it was possible to release nuclear energy.

Uranium

The element uranium, chemical symbol U, is named for the planet Uranus, since the planet was discovered in 1781, just eight years before the element uranium.

Uranium, a silvery-white, extremely dense metal, was first discovered in the mineral pitchblende. Uranium is not an especially rare mineral. It is more plentiful than, say, mercury or silver. However, it has become of vital importance in the nuclear age.

An overwhelmingly large proportion of uranium on Earth is uranium-238. This makes it the heaviest atom commonly found in nature. Uranium turns blue in air because it develops an oxide coating.

Uranium is not found in concentrated form; many tonnes of ore have to be processed to obtain even one gram of the element.

The biggest deposits of uranium ore are at Blind River, Canada, in South Africa, Australia, France, and in Colorado and Utah in the United States.

The most important compound of uranium is uranium hexafluoride gas, which can be used to separate uranium-238 from uranium-235, the main ingredient of the atomic bomb.

Release of energy

Uranium can be used to release enormous amounts of energy, for both peaceful and military purposes. The energy released by a kilogram of uranium is about two and a half million times the energy released by burning one kilogram of coal.

The main way the energy is released is known as nuclear fission. It is described on page 28.

▶ A uranium mine in Namibia. Notice how the mine is geared to extracting very large quantities of rock. The structures on the site include concentrating plants to improve the proportion of element to rock as much as possible before transporting it to a smelter.

▼ A sample of uranium ore or pitchblende.

Uranium reprocessing

Concentrated "enriched" uranium oxide (containing more uranium-235 than normal) is used in the form of rods as a fuel in nuclear reactors. Over time the reaction inside the nuclear reactor changes the uranium to other materials. Thus the amount of uranium left in the fuel rods decreases, and a point is reached where there is not enough uranium left to keep the reactor going. At this stage the uranium has to be reprocessed, separating out the unspent uranium and reforming it into rods.

Uranium produces helium

As uranium decays it releases alpha particles, which capture electrons from other atoms to form helium gas. This is the source of helium gas trapped in ancient rocks. However, uranium can also be made to split into new nuclei when it is bombarded by neutrons. This latter property is used in weapons and reactors.

Other common radioactive elements

A whole series of radioactive elements has been discovered in the last century. Most of these do not occur naturally, or, if they do, exist for only very short periods of time.

In fact, many new elements can be made in the laboratory by bombarding an element with neutrons. This is done inside a nuclear reactor.

Plutonium

Plutonium, a silvery metal with a half-life of 24,000 years, was discovered in 1940 and named after the planet Pluto. It occurs in very tiny quantities in association with uranium. Large lumps of plutonium are warm to the touch due to the natural release of energy within the material. Most of the plutonium needed is made from uranium.

Plutonium is a vital fuel for nuclear reactors and nuclear bombs. It is used especially as a fuel in the types of reactor designed by France and Russia. A mass of just 5 kg of plutonium (about the size of a large orange) will make a nuclear bomb.

Plutonium has also been used successfully to make long-life batteries used in heart pacemakers and for powering long-distance spacecraft such as the Voyager series.

The radiation from a piece of plutonium is very readily shielded, for example, by rubber gloves. The major concern is from plutonium dust that might be inhaled. Plutonium is chemically similar to calcium and therefore can replace it in bones, irradiating the nearby marrow cells and causing leukaemia.

► The fuel element from an advanced gas-cooled reactor. When filled with fuel, one of these elements produces as much energy as about 3000 tonnes of coal.

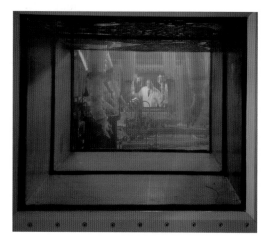

◄ Handling radioactive materials in lead-lined chambers.

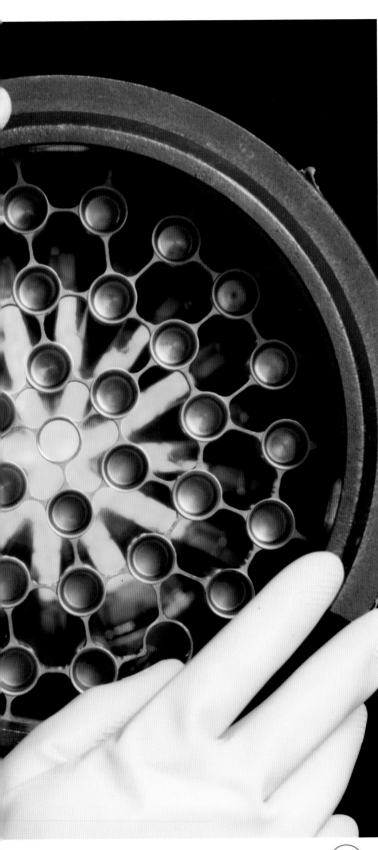

Strontium

Strontium is a soft, silvery metal that behaves very much like calcium. Strontium-90 has a half-life of 28 years.

Strontium was discovered nearly two centuries ago, but its radioactive forms were only discovered much more recently.

Strontium-90 is one of the products of nuclear explosions, and as a result of them, levels of strontium in the atmosphere increased during periods of atmospheric nuclear tests.

Because it becomes absorbed by the bones and can irradiate marrow cells and cause leukaemia, strontium is thought to be the most dangerous form of radioactive fallout from nuclear explosions.

Cobalt

Cobalt (symbol Co) is a hard, silver metal. It gets its name from the German *Kobold* (a mythological underground demon). The ores in which cobalt are found also contain arsenic, and the first people to work the ore died from arsenic poisoning. Ironically, cobalt is a vital trace element and important for health. Zaïre has two-thirds of the world's reserves of cobalt.

Cobalt-60 is an artificially produced isotope (form) of cobalt. It is made by bombarding naturally occurring cobalt-59 with neutrons inside an atomic reactor. Cobalt-60 has healing properties when used as part of radiation therapy. It is used as a source of gamma radiation, which is similar to X-rays.

Fission: splitting the atom

Fission is the process of splitting up the core, or nucleus, of an atom into two smaller pieces.

When an atom breaks apart, each of the parts carries with it more energy than it would have in nature, and this energy is lost as heat, radiation and particles.

It is easier to get some materials to split apart than others. Uranium is the easiest, which explains its widespread use in the nuclear industry.

When enough uranium is brought together, the release of neutrons by spontaneous decay is sufficient for some neutrons to be captured by other nuclei, splitting them and so starting a "chain reaction". Controlling this chain reaction controls the output of the reactor in a nuclear power station; letting the reaction continue at an uncontrolled rate produces a nuclear explosion (see page 32).

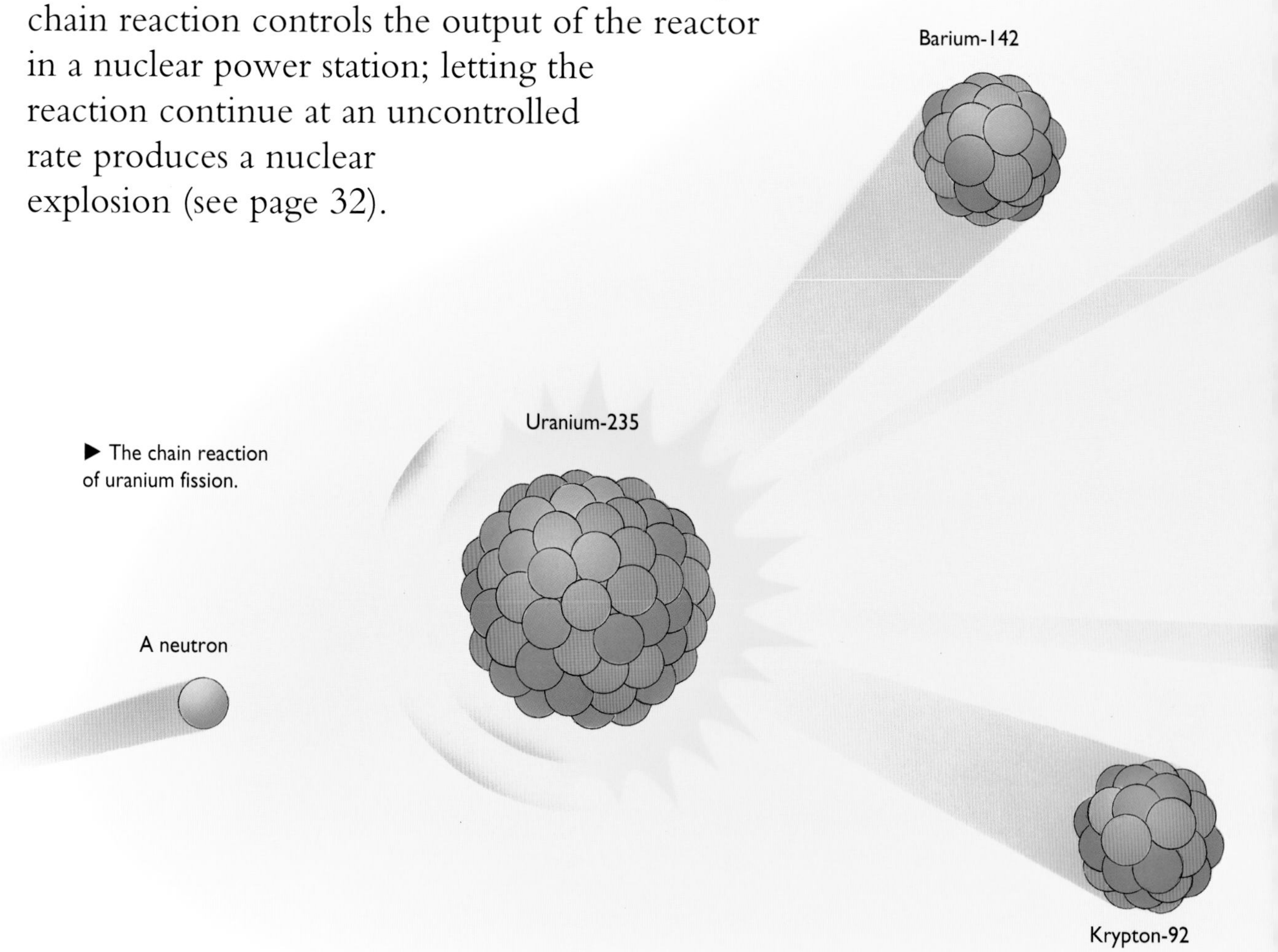

Barium-142

Uranium-235

▶ The chain reaction of uranium fission.

A neutron

Krypton-92

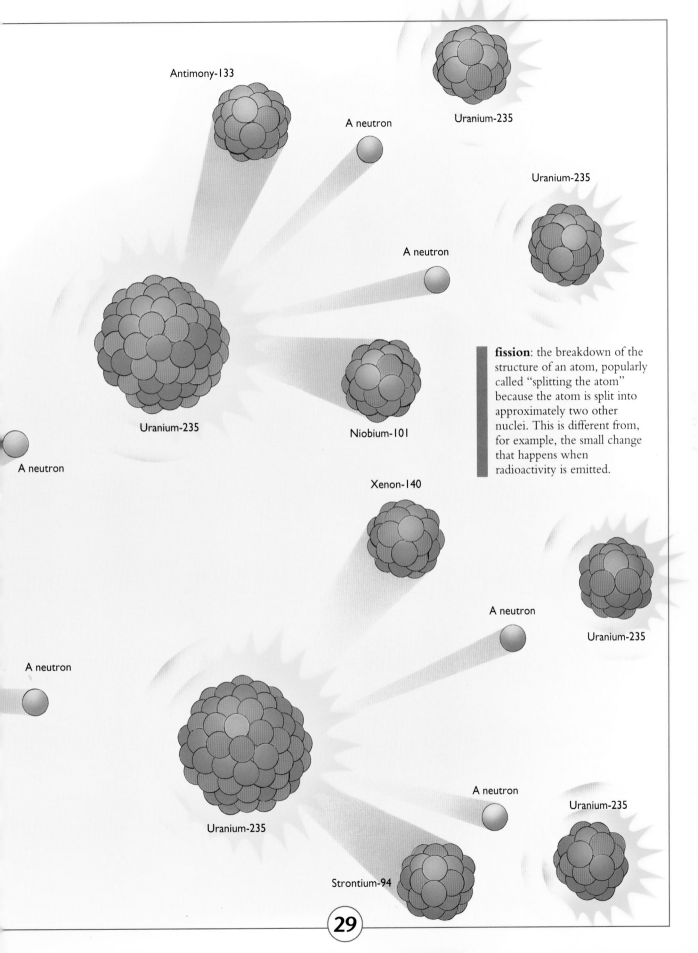

Antimony-133

A neutron

Uranium-235

Uranium-235

A neutron

Uranium-235

Niobium-101

fission: the breakdown of the structure of an atom, popularly called "splitting the atom" because the atom is split into approximately two other nuclei. This is different from, for example, the small change that happens when radioactivity is emitted.

Xenon-140

A neutron

A neutron

Uranium-235

A neutron

A neutron

Uranium-235

Uranium-235

Strontium-94

Fusion: combining atoms

Whereas nuclear fission is splitting the atom, nuclear fusion forces two atoms to come together, or fuse, to produce an even heavier atom. The fusion of atoms releases huge amounts of energy.

Fusion has proved to be one of the most difficult tasks for scientists to perform, yet it is one of the commonest features of the Universe. The reason scientists find fusion so difficult is that enormously high temperatures and great pressures are needed to cause the reaction. This commonly happens in the stars. So getting fusion going on Earth is nothing short of making stars!

From hydrogen to helium

The main reaction that occurs inside a star is the fusion of the simplest atom – hydrogen – into the next simplest atom, helium. This happens under the pressure created by the huge amount of gravity created in a star. Inside a star atoms are literally pulled together by gravity. As they fuse, they release energy, which causes the star to rise in temperature until it may be well over 10 million degrees. This is called a thermonuclear reaction.

An endless supply of cheap energy

Scientists know that if they can use the fusion process to make energy, they can use hydrogen from ordinary water (in a form called heavy water). This supply will last for billions of years.

However, because a nuclear reactor can only be a tiny fraction of the size of a star, the size (and gravity) must be compensated for by creating even higher temperatures than exist in stars. Scientists are working to solve this problem today, and it is likely they will achieve success by the middle of the twenty-first century.

▶ The Sun is made mainly of hydrogen. Its immense gravitational power pulls the hydrogen atoms together to form helium. The flow of energy so produced makes the Sun extremely hot, and enormous amounts of energy are released to space. A small fraction of this energy reaches Earth as sunlight, ultraviolet light and infra-red light. In this way nuclear fusion reactions already make life on Earth possible.

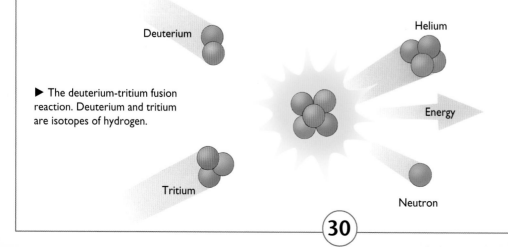

Deuterium

Helium

▶ The deuterium-tritium fusion reaction. Deuterium and tritium are isotopes of hydrogen.

Energy

Tritium

Neutron

fusion: combining atoms to form a heavier atom.

thermonuclear reactions: reactions that occur within atoms due to fusion, releasing an immensely concentrated amount of energy.

▼ The elements are made inside the stars. Different elements form at different stages of the life cycle of a star. Hydrogen burns to form helium in the young bright stars; helium reacts to yield carbon. This process then continues as stars change and become red giants in which nitrogen, oxygen, iron, cobalt and nickel are formed.

As the stars finally explode to form the huge gas clouds called supernovas, fusion creates the heavier elements before the gas loses its heat to the immensity of space.

Pollution-free radiation

The real benefit of fusion is that it does not create any radioactive pollution. Fusion is not a chain reaction like fission and so cannot run out of control and cause a massive explosion. Thus, the present generation of nuclear reactors is just a temporary step on the way to a nuclear energy goal. This is why countries need to pursue nuclear technology, so that they do not fall behind and can really benefit from the development of fusion power when it comes.

Nuclear energy

The most common use of the radioactive elements is to release nuclear energy. Radioactive elements are not, of course, alone in releasing energy as they change: all matter releases energy. If we eat a bar of chocolate, for example, chemical changes occur that release energy to our bodies. Another familiar example is the way natural gas releases energy as it combines with oxygen during burning.

In a nuclear reaction the change takes place *inside* the atoms. A radioactive material such as uranium is bombarded with radiation in the form of neutrons. This process, fission, causes the uranium to split up into two new atoms, setting in motion a self-bombarding process that releases massive amounts of heat.

Uranium bombarded with neutrons produces strontium and xenon gas and releases more neutrons, giving out more heat energy.

In this case the heat given out is twenty-five million times as great as for an equivalent amount of burning gas.

▼ Countries which have few natural reserves of fossil fuels tend to invest in nuclear energy to meet their electricity needs. In Japan half of all electricity is produced by nuclear power; in France it is three-quarters.

▶ A nuclear power station, showing the reactor in the foreground and cooling towers to the left.

Uses of nuclear energy

The image of nuclear energy is clouded with the experience of how the energy can be released for destructive purposes (such as the atomic bombs dropped on Japan in order to end World War II, see page 38) and how some dangerous radioactive materials can sometimes escape (as was the case at Chernobyl, see page 37). Indeed, the need for very powerful bombs during World War II encouraged politicians to fund nuclear research in the first place.

The first fission reactor was created by a team led by Enrico Fermi in Chicago in 1942. His team used natural uranium, which is mostly uranium-238. However, only the more rare uranium-235 can be made to react. Thus, they discovered that to make the reaction work successfully they needed to produce concentrated uranium-235 ("enriched uranium"), or they needed to use plutonium.

Although some of this material has been used for weapons, the vast

majority is now used for making electricity. Nearly one-fifth of the world's electricity is generated by nuclear means, making it one of the world's most important energy supplies. It is expected that although nuclear energy programmes have slowed down in some countries, the adoption of nuclear energy in the developing world – especially in countries that have little coal or oil of their own – will cause the world use of nuclear energy to be responsible for one-quarter of the world's electricity needs by the end of the century.

▲ The core of a reactor contains fuel rods, a moderator and control rods. The purpose of the fuel rods is to provide the fissionable material. However, without a moderator material (such as graphite), the neutrons created by fission would move too fast to be captured by the remaining fuel. The moderator slows down the neutrons so that the likelihood of capture by a nucleus is high. Control rods, on the other hand, serve to stop the fission process. They are made of materials that absorb (not just slow down) neutrons, so that the neutrons cannot be captured by more nuclei.

Reactor

Nuclear reactors

A nuclear reactor is a form of power unit used to generate electrical energy. It is flexible enough to be able to supply electricity for use in homes and factories, or to power submarines and warships.

Nuclear reactors make their energy from the process of fission (see page 28), using uranium and plutonium as fuels. The fission process is started when the fuel rods are bombarded with neutrons. Once the reaction is started, it develops a chain reaction that could cause a meltdown if not controlled. Many of the main features of a nuclear power plant are designed simply to control the chain reaction.

Warship reactors
It has become common for nuclear power plants to be used on military vessels, especially submarines and aircraft carriers. This gives them exceptional range and removes the need for refuelling (new fuel rods have to be fitted only after more than a million km!).

In the United States fleet, some carriers such as the *Enterprise* have eight nuclear reactors. Most in the *Nimitz* class have two reactors.

These vast ships come with a heavy price tag. The *Theodore Roosevelt*, for example, cost $17 billion after it had been fitted with aircraft and provided with escort vessels.

Types of reactors

The main parts of a reactor are the same worldwide, although different cooling fluids are used in different designs. Commonly, water or gas is used as a coolant to transfer the heat energy from the reactor to the turbines of the generating station and to prevent the reactions from getting out of control.

At the heart of the reactor is the fuel, usually uranium oxide or plutonium. This is made into the form of about 200 three-metre-long fuel rods that can be lowered or raised from a surrounding fluid (water or carbon dioxide gas). Along with them may be rods made of steel alloys that absorb neutrons. Cadmium is commonly used as an alloying metal.

The control rods are raised or lowered among the fuel rods to speed up, slow down or stop the chain reaction altogether. Raising and lowering the absorbing rods is the same as stepping on or lifting your foot off the accelerator in a vehicle. In some types of reactor water is used to control the reaction as well as to carry away the heat.

The fuel rods and control rods may be packaged in a fuel cell about four metres across and four metres high. They have to be sealed inside a container that is designed to stand up to high pressures, high temperatures and explosions. It is normally made of very thick steel and may be twelve metres high. This steel vessel contains the cooling fluid that soaks up the heat from the reactor and transfers it to the turbines.

The cooling fluid becomes radioactive during use. It cannot therefore be fed directly to the turbines. Instead a set of pipes is wrapped around the cooling fluid pipes, and heat is exchanged. This fluid does not become radioactive and so is safe for use in the generator.

All of the parts of the reactor and its cooling fluid are housed in a large concrete building designed to contain any explosion or other misadventure. This concrete building is normally shaped in the form of a large ball or a "bullet" standing on end. Thus, when you see a nuclear power plant, you can tell where the reactor is housed by looking for the round-shaped buildings.

Breeder reactor

It is possible for a nuclear reactor to be designed to make more fuel than it uses, considerably cutting down fuel costs.

Breeder reactors consist of many stainless steel tubes containing uranium oxide and plutonium oxide, surrounded by tubes containing uranium oxide. As the fuel in the central rods undergoes fission, it releases neutrons that are absorbed by the uranium in the outer rods. These convert the uranium in the outer rods into plutonium. The better use of uranium makes the fuel last up to fifty times as long as in ordinary reactors.

Normally the heat produced by the fission of uranium and plutonium in the core is used to heat water, which is then taken away to drive steam turbines and make electricity. But the water also soaks up neutrons. To allow the neutrons to escape from the core and produce new plutonium, a different material has to be used. In fast breeder reactors the water is replaced with liquid sodium.

alloy: a mixture of a metal and various other elements.

fission: the breakdown of the structure of an atom, popularly called "splitting the atom".

Using uranium in nuclear reactors

The element uranium is most widely used as the fuel of a nuclear reactor. In a nuclear reactor the fission process is started by spontaneous decay because of so much fuel being contained together. This sets up a chain reaction that can be controlled so that new reactions occur all the time. The rate of the reactions determines the amount of heat released and thus eventually the amount of electricity that the power stations can produce.

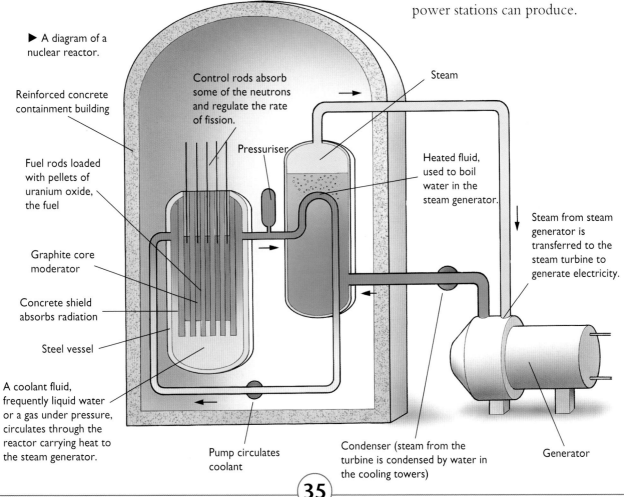

▶ A diagram of a nuclear reactor.

Reinforced concrete containment building

Control rods absorb some of the neutrons and regulate the rate of fission.

Steam

Fuel rods loaded with pellets of uranium oxide, the fuel

Pressuriser

Heated fluid, used to boil water in the steam generator.

Graphite core moderator

Steam from steam generator is transferred to the steam turbine to generate electricity.

Concrete shield absorbs radiation

Steel vessel

A coolant fluid, frequently liquid water or a gas under pressure, circulates through the reactor carrying heat to the steam generator.

Pump circulates coolant

Condenser (steam from the turbine is condensed by water in the cooling towers)

Generator

Nuclear accidents

A nuclear reactor produces the most concentrated radioactive materials in the world, and special precautions need to be taken to ensure safety.

The greatest danger is that the core may become overheated. This might happen if the cooling fluid is not circulating properly or if the control rods or control fluid are not adjusted properly so that too much power is released from the fuel rods. If the reactor gets too hot, the cooling fluid may boil. The bubbles in the cooling fluid are not effective at carrying away heat, so the reactor can then get so hot it begins to melt. Boiling fluid and lack of cooling water were the main causes of the disaster at Chernobyl (see opposite).

To prevent this scenario there are many safety devices that sense increases in temperature and shut the reactor down slowly and safely, as appropriate.

Three Mile Island

A number of releases of plutonium have given rise to great concern. One of the most widely reported occurred after an accident in 1979 on Three Mile Island near Harrisburg, Pennsylvania, where a nuclear power station accident caused small amounts of plutonium to be released into the air and the water. The plant that caused the trouble has now been sealed up.

Radioactive watches

Women employed during the 1920s as painters of luminescent clock dials were unwittingly exposed to radium from the paint as they licked their brushes. Many died, either from anaemia or from bone cancer, alerting doctors to the dangers of radioactivity and subsequent radiation injury. Other types of life-threatening bone disease can be caused by overexposure to X-rays.

▲▼ The crippled reactor at Chernobyl.

Chernobyl

The word Chernobyl is familiar to almost everyone worldwide. This notoriety began on April 26, 1986, when one of the reactors at the Chernobyl site in the Ukraine exploded.

The explosion was the result of poor workmanship during construction, poorly designed safety systems and a series of disastrous decisions by the people operating the plant. In fact the reactor was being used for experiments and its emergency cooling system had been shut down. The experiment went wrong and the reactor began to heat up out of control. There was no cooling system to deal with the problem, and eventually an explosion occurred that blew the lid off the reactor. (It is important to note that, at the time, Soviet designs did not include the ball-shaped containment vessel used today to contain the results of any explosion. This made the Soviet design far cheaper but also far more dangerous.)

The reactor fuel started to melt its way down through the floor of the reactor. At the same time, not only was highly radioactive material thrown out of the top of the reactor, but far worse, there was a huge escape of radioactive gas and small particles (called contaminants). These went into the atmosphere where they were carried for thousands of kilometres by air currents.

As the contaminants were washed from the sky by rain, they settled out on the plants of the land over much of northern Europe, contaminating the plants and making them unfit for eating. Animals grazing the grass in the contaminated area then produced contaminated milk, and their meat became unfit for consumption.

After some days, and through a number of daring and dangerous manoeuvres, the reactor was encased in concrete and its radioactivity contained.

The disaster killed 31 people outright, but the real death toll, including those who died or will die as a result of being exposed to radiation – perhaps within a 100 km radius of the plant – will run into the thousands.

The Chernobyl disaster has made people worry about nuclear power worldwide, even though nuclear power stations in most countries are built to much higher standards, and such accidents probably would not happen to them.

Nuclear weapons

Everyone knows that the most powerful weapons in the world are nuclear weapons. The first nuclear bomb was the atomic bomb.

Fission bombs

The atomic bomb releases its energy through an uncontrolled chain reaction of the same kind that is used in nuclear power stations. That is, the atomic bomb is a device for splitting atoms (see page 28).

The atomic bomb was developed in the United States during World War II. The development of the bomb was called the Manhattan Project, and it used many of the world's top scientists. It was begun because it was feared that the scientists in Hitler's Germany were making such a bomb, and it was essential that the allies have a weapon to counter it.

During this time an enormous amount was learned about how atoms work; and this, in the long run, was to prove very useful for making nuclear reactors for peaceful purposes. For the bombs scientists enriched the uranium to about 90% (it is a mere 3% in power stations.).

The bomb was made of two parts, each containing a piece of uranium too small to produce nuclear fission on its own. When the bomb fell on Hiroshima, Japan, on August 6, 1945, a high explosive charge forced these two pieces of uranium together very rapidly and held them in place for the microseconds needed for the uncontrolled chain reaction to occur.

Plutonium can also be used to split the atom. This was the material used for the bomb dropped on Nagasaki on August 9, 1945.

Fusion bombs

The fusion reaction is far more powerful than the fission reaction, but it is also harder to create. The hydrogen bomb, H-bomb, or thermonuclear device, was the first nuclear bomb to use fusion. An H-bomb is about a thousand times more powerful than an atom bomb (A-bomb) for the same mass of materials.

The problem with a fusion reaction is that it will only work with extraordinarily high temperatures. This is provided by exploding an A-bomb next to the materials that will be involved in fusion. Thus the A-bomb becomes the trigger for the fusion bomb. Today all large nuclear devices use fusion as the means of producing an explosion.

Various isotopes of hydrogen, including deuterium (which can be obtained from "heavy water", D_2O) and tritium, are now used for the H-bomb.

A 20 megaton bomb can probably destroy everything within a 16 km radius of where it falls. This devastating effect is too ghastly to contemplate, which is why most major nuclear countries continue to make efforts to reduce their stockpiles of such weapons, and why they try to stop other countries from making such weapons.

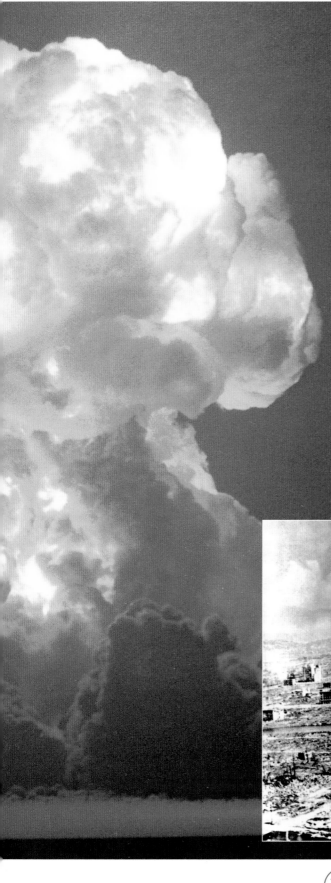

Fallout

Fallout is a word used for the way that nuclear materials released into the atmosphere gradually find their way back to the ground.

Natural fallout occurs all the time due to the way that radioactive materials (for example, carbon-14) are created constantly as the atmosphere is bombarded by cosmic rays. Other local, but extremely intense fallout, is produced from nuclear explosions or nuclear reactor accidents.

To understand how fallout particles may contaminate an area, you need to understand how the atmosphere works. Three kinds of fallout can occur:

(i) Close to the release of radioactive material, the fallout of larger particles will occur due to their own weight. This will be severe but will not last a long time.

(ii) Smaller particles released into the air will be carried aloft by winds, where they may be combined with water droplets and fall out as rain over a wide area. Other particles that do not combine with water droplets may remain in the air for long periods of time, possibly months.

(iii) Large nuclear explosions send a column of expanding gas that punches a hole through the lower part of the atmosphere and sends clouds of radioactive particles into the upper air, the region called the stratosphere. There is no cloud in this region to help wash the particles from the air, so they may spread around the globe, remaining in the air for many years. The main radioactive elements in this high level fallout are strontium-90 and cesium-137.

◀ A nuclear bomb test.

▼ On August 9, 1945, United States forces released the second of the two atomic bombs dropped on Japan. The attack killed 73,884 people and destroyed 47% of the city of Nagasaki.

Nuclear waste

As scientists use the nuclear materials they need, some waste material is produced, much of which is radioactive. More radioactive material comes from the spent fuel rods in reactors.

The key to nuclear reactions is that they happen inside each atom. We have no means to stop these changes once they have begun, so radioactive materials must simply be allowed to lose their radioactivity naturally. Materials like uranium, with half-lives of billions of years, will never become safe.

The danger is that the radiation emitted from waste material can cause damage to human tissue, so people must be shielded from all sources of such radiation.

High level waste

Each power station produces about the same volume of high level waste as the size of an automobile each year.

Although the volume is small, high level waste such as spent fuel rods from power stations and discarded nuclear warheads causes many problems. There is no choice but to encase the material in glass (which does not corrode) and bury it in very deep mines. Fortunately, at the moment the volume of such waste is not large, and it is likely that as fusion becomes a possible way of generating electricity, the amount of new radioactive waste will be much smaller than in the past. Scientists therefore have to cope with a problem that may be at its worst for the next half century.

▼ Burial of intermediate level waste.

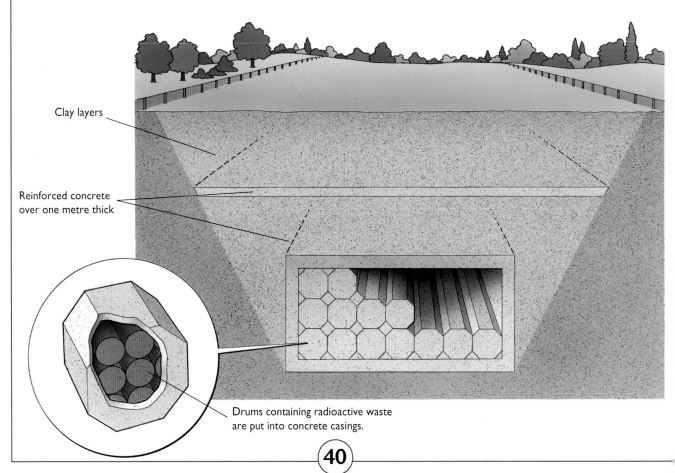

Clay layers

Reinforced concrete over one metre thick

Drums containing radioactive waste are put into concrete casings.

Low level waste
Some waste is not very radioactive. This includes the waste rock from uranium mines and the discarded clothing from scientists working with radioactive materials, radiologists in hospitals, etc. This material is normally buried in special dumps. A covering of soil is sufficient to act as a shield.

▲ Decommissioning nuclear reactors is a very difficult process because the amount of high level waste is large. In many cases the reactor core will be left where it is and buried in an immense slab of concrete.

▶ Low level waste can be placed in steel and concrete drums and taken to suitable land or ocean dumps.

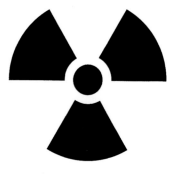

▲ Radiation symbol.

Irradiation

The word radiation has come to mean the harmful effects that radioactivity can have on the body. However, radiation can be a very useful process. Even despite accidents in the past, its usefulness far outweighs its dangers.

The key to understanding how to make use of radiation, and the precautions to take, is to understand what radiation does to living tissue.

Effects of exposure to radiation

Alpha rays are made of very bulky particles that eventually combine with electrons from other atoms to produce helium gas. They do not travel far inside the body. But the alpha particles cause great damage to body tissues because they can knock atoms out of molecules and cause them to change or mutate. The effect is especially severe if the DNA inside cells is damaged. The result may be that the wrong signals are sent to build new tissue, and cancers may be formed.

Most body damage from alpha particles is caused to the DNA in skin cells. Beta radiation penetrates tissue more deeply than alpha radiation, reaching 2-3 cm inside the tissues. However, the most powerful form of radiation is gamma radiation, similar to X-rays. These waves of nuclear energy can pass right through the body and so can damage any atoms anywhere. Severe doses of this form of radiation can affect any of the body's inside organs, causing what is known as "radiation sickness".

Whichever form of radiation is involved, an overdose causes many cells to die or mutate. A general name for mutated cells is cancer, and natural radiation has been one contributor to many skin cancers.

Atom bomb testing in the atmosphere produced radioactive strontium-90. Strontium behaves chemically very much like calcium and is easily incorporated into the bones. It is believed that strontium in bone cells causes damage to blood cells, a form of cancer known as leukaemia. For this reason, there have been no atmospheric nuclear tests for many years.

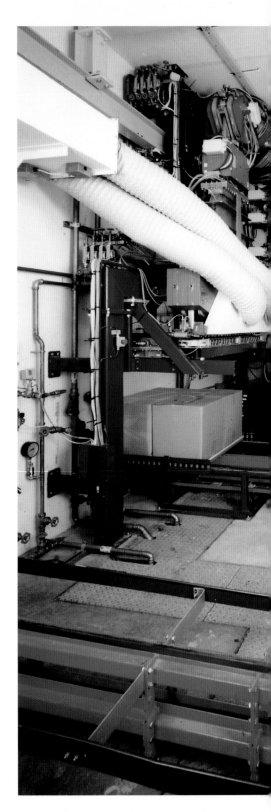

▲ The boxes on this conveyor belt are being irradiated to sterilise the medical equipment they contain.

Measuring radiation doses

Scientists measure the radiation received with a unit called the rem (roentgen equivalent in man). A radiation level of 10,000 rem will disrupt the central nervous system so much that people will die of radiation. Even a dose of 300 rem will kill half of the people exposed to it, while anything over 100 rem will most likely cause some form of injury.

If irradiation affects the cells of the bone, then enough cells may die to cause a loss of the natural ability of the body to resist infection. If the radiation affects the eyes, it may cause cataracts and loss of vision. It may also cause sterility.

It is difficult to determine the risks of low level radiation; even a body scanner will produce a rem of 0.1. This is why it is important to protect people working with radiation, allowing them to work behind lead or thick concrete shields.

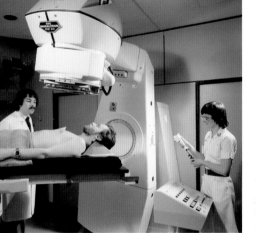

▲ A patient receiving radiation therapy.

▲ One way in which radiation can be used beneficially is to irradiate food. By subjecting food to massive doses of radiation, all harmful organisms are killed. Irradiation is therefore a very thorough technique of sterilisation. Some foods found in grocery stores are now sterilised by irradiation. The packaging should carry the symbol shown above.

Radiation therapy

Radiation therapy uses radiation to treat illness. In its use for cancer treatment, the patient is exposed to radiation from a cobalt-60 or iodine-131 source.

The amount of chosen doses of radiation is designed to kill the malignant (actively changing) cells without harming the rest of the body. This is possible because malignant tissues are more sensitive to radiation than healthy tissues.

The way the radiation is applied is normally from an external "cannon" that "fires" a high-energy beam which is focused at the target. The normal treatment is a few minutes a day for one or two months.

Key facts about...

Uranium

A soft, silvery metal,
chemical symbol U

Harmful to life

Decays naturally
through a radioactivity
series to lead

Heavy element

Melting point 1132°C

Used as a source
of fissionable fuel

Poor conductor
of electricity

Dissolves in acids

Atomic number 92,
atomic weight about 238

SHELL DIAGRAMS

The shell diagram on this page
is a representation of an atom
of the element uranium. The total
number of electrons is shown in
the relevant orbitals, or shells,
around the central nucleus.

Electron shell

Electron

Nucleus containing
protons and neutrons
(called nucleons)

▶ Advanced gas-cooled
reactor fuel elements being
placed in a reactor core.

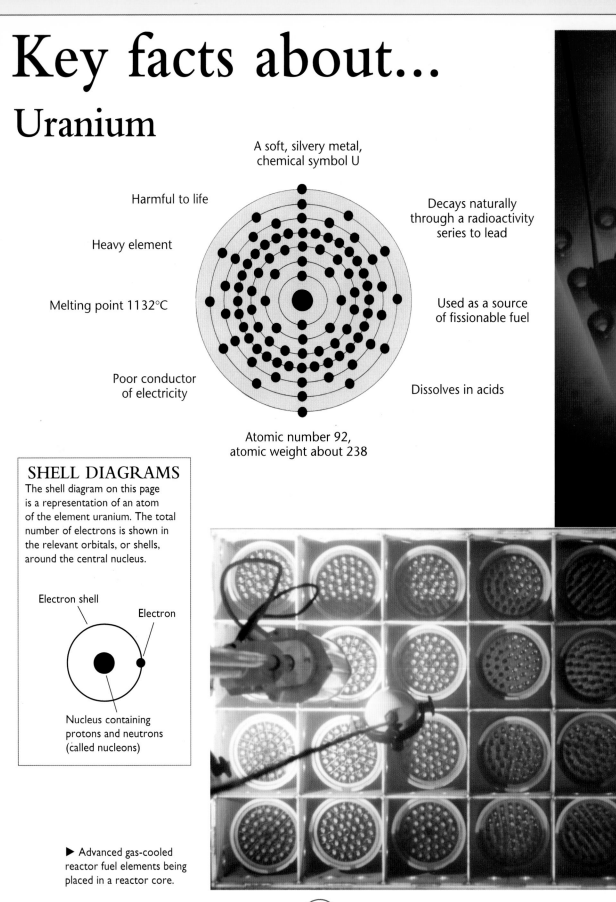

▲ Cerenkov radiation causes the water pond containing these spent fuel rods from a nuclear reactor to glow blue.

Cerenkov radiation may be thought of as a shock wave, much like a sonic boom from an aircraft travelling faster than the speed of sound. The Cerenkov effect occurs because atomic particles travel through water faster than light travels through water.

◄ Reprocessing a uranium billet.

The Periodic Table

The Periodic Table sets out the relationships among the elements of the Universe. According to the Periodic Table, certain elements fall into groups. The pattern of these groups has, in the past, allowed scientists to predict elements that had not at that time been discovered. It can still be used today to predict the properties of unfamiliar elements.

The Periodic Table was first described by a Russian teacher, Dmitry Ivanovich Mendeleev, between 1869 and 1870. He was interested in writing a chemistry textbook, and wanted to show his students that there were certain patterns in the elements that had been discovered. So he set out the elements (of which there were 57 at the time) according to their known properties. On the assumption that there was pattern to the elements, he left blank spaces where elements seemed to be missing. Using this first version of the Periodic Table, he was able to predict in detail the chemical and physical properties of elements that had not yet been discovered. Other scientists began to look for the missing elements, and they soon found them.

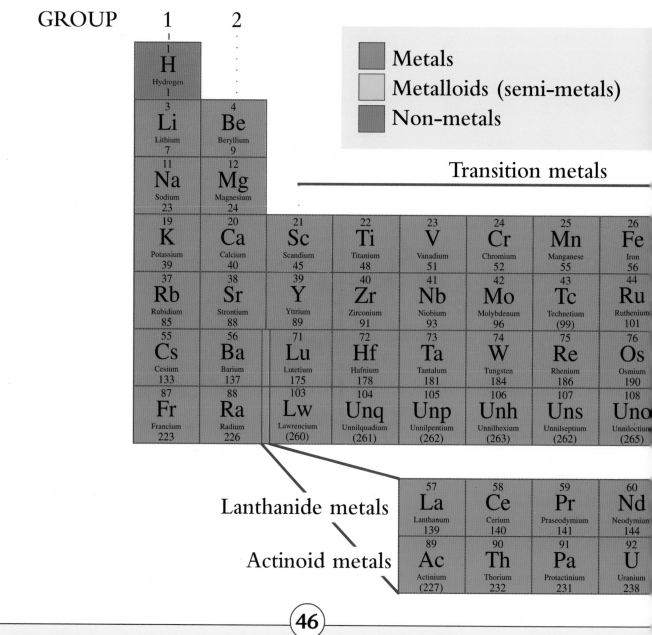

Hydrogen did not seem to fit into the table, so he placed it in a box on its own. Otherwise the elements were all placed horizontally. When an element was reached with properties similar to the first one in the top row, a second row was started. By following this rule, similarities among the elements can be found by reading up and down. By reading across the rows, the elements progressively increase their atomic number. This number indicates the number of positively charged particles (protons) in the nucleus of each atom. This is also the number of negatively charged particles (electrons) in the atom.

The chemical properties of an element depend on the number of electrons in the outermost shell.

Atoms can form compounds by sharing electrons in their outermost shells. This explains why atoms with a full set of electrons (like helium, an inert gas) are unreactive, whereas atoms with an incomplete electron shell (such as chlorine) are very reactive. Elements can also combine by the complete transfer of electrons from metals to non-metals and the compounds formed contain ions.

Radioactive elements lose particles from their nucleus and electrons from their surrounding shells. As a result their atomic number changes and they become new elements.

Atomic (proton) number — 13
Symbol
Al
Aluminium — Name
27
Approximate relative atomic mass (Approximate atomic weight)

3	4	5	6	7	0
					2 He Helium 4
5 B Boron 11	6 C Carbon 12	7 N Nitrogen 14	8 O Oxygen 16	9 F Fluorine 19	10 Ne Neon 20
13 Al Aluminium 27	14 Si Silicon 28	15 P Phosphorus 31	16 S Sulphur 32	17 Cl Chlorine 35	18 Ar Argon 40

27 Co Cobalt 59	28 Ni Nickel 59	29 Cu Copper 64	30 Zn Zinc 65	31 Ga Gallium 70	32 Ge Germanium 73	33 As Arsenic 75	34 Se Selenium 79	35 Br Bromine 80	36 Kr Krypton 84
45 Rh Rhodium 103	46 Pd Palladium 106	47 Ag Silver 108	48 Cd Cadmium 112	49 In Indium 115	50 Sn Tin 119	51 Sb Antimony 122	52 Te Tellurium 128	53 I Iodine 127	54 Xe Xenon 131
77 Ir Iridium 192	78 Pt Platinum 195	79 Au Gold 197	80 Hg Mercury 201	81 Tl Thallium 204	82 Pb Lead 207	83 Bi Bismuth 209	84 Po Polonium (209)	85 At Astatine (210)	86 Rn Radon (222)

109 Une Unnilennium (266)

61 Pm Promethium (145)	62 Sm Samarium 150	63 Eu Europium 152	64 Gd Gadolinium 157	65 Tb Terbium 159	66 Dy Dysprosium 163	67 Ho Holmium 165	68 Er Erbium 167	69 Tm Thulium 169	70 Yb Ytterbium 173
93 Np Neptunium (237)	94 Pu Plutonium (244)	95 Am Americium (243)	96 Cm Curium (247)	97 Bk Berkelium (247)	98 Cf Californium (251)	99 Es Einsteinium (252)	100 Fm Fermium (257)	101 Md Mendelevium (258)	102 No Nobelium (259)

Understanding equations

As you read through this book, you will notice that many pages contain equations using symbols. If you are not familiar with these symbols, read this page. Symbols make it easy for chemists to write out the reactions that are occurring in a way that allows a better understanding of the processes involved.

Symbols for the elements

The basis of the modern use of symbols for elements dates back to the 19th century. At this time a shorthand was developed using the first letter of the element wherever possible. Thus "O" stands for oxygen, "H" stands for hydrogen

and so on. However, if we were to use only the first letter, then there could be some confusion. For example, nitrogen and nickel would both use the symbols N. To overcome this problem, many elements are symbolised using the first two letters of their full name, and the second letter is lowercase. Thus although nitrogen is N, nickel becomes Ni. Not all symbols come from the English name; many use the Latin name instead. This is why, for example, gold is not G but Au (for the Latin *aurum*) and sodium has the symbol Na, from the Latin *natrium*.

Compounds of elements are made by combining letters. Thus the molecule carbon

Written and symbolic equations

In this book, important chemical equations are briefly stated in words (these are called word equations), and are then shown in their symbolic form along with the states.

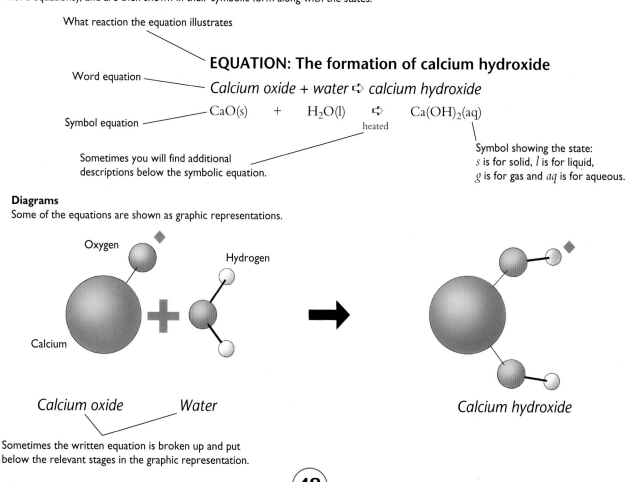

What reaction the equation illustrates

EQUATION: The formation of calcium hydroxide

Word equation —— *Calcium oxide + water ⇨ calcium hydroxide*

Symbol equation —— $CaO(s)$ + $H_2O(l)$ ⇨ $Ca(OH)_2(aq)$
heated

Sometimes you will find additional descriptions below the symbolic equation.

Symbol showing the state:
s is for solid, l is for liquid,
g is for gas and aq is for aqueous.

Diagrams

Some of the equations are shown as graphic representations.

Oxygen

Hydrogen

Calcium

Calcium oxide   *Water*

Calcium hydroxide

Sometimes the written equation is broken up and put below the relevant stages in the graphic representation.

monoxide is CO. By using lowercase letters for the second letter of an element, it is possible to show that cobalt, symbol Co, is not the same as the molecule carbon monoxide, CO.

However, the letters can be made to do much more than this. In many molecules, atoms combine in unequal numbers. So, for example, carbon dioxide has one atom of carbon for every two of oxygen. This is shown by using the number 2 beside the oxygen, and the symbol becomes CO_2.

In practice, some groups of atoms combine as a unit with other substances. Thus, for example, calcium bicarbonate (one of the compounds used in some antacid pills) is written $Ca(HCO_3)_2$. This shows that the part of the substance inside the brackets reacts as a unit and the "2" outside the brackets shows the presence of two such units.

Some substances attract water molecules to themselves. To show this a dot is used. Thus the blue form of copper sulphate is written $CuSO_4.5H_2O$. In this case five molecules of water attract to one of copper sulphate.

When you see the dot, you know that this water can be driven off by heating; it is part of the crystal structure.

In a reaction substances change by rearranging the combinations of atoms. The way they change is shown by using the chemical symbols, placing those that will react (the starting materials, or reactants) on the left and the products of the reaction on the right. Between the two, chemists use an arrow to show which way the reaction is occurring.

It is possible to describe a reaction in words. This gives word equations, which are given throughout this book. However, it is easier to understand what is happening by using an equation containing symbols. These are also given in many places. They are not given when the equations are very complex.

In any equation both sides balance; that is, there must be an equal number of like atoms on both sides of the arrow. When you try to write down reactions, you, too, must balance your equation; you cannot have a few atoms left over at the end!

The symbols in brackets are abbreviations for the physical state of each substance taking part, so that (*s*) is used for solid, (*l*) for liquid, (*g*) for gas and (*aq*) for an aqueous solution, that is, a solution of a substance dissolved in water.

Atoms and ions
Each sphere represents a particle of an element. A particle can be an atom or an ion. Each atom or ion is associated with other atoms or ions through bonds – forces of attraction. The size of the particles and the nature of the bonds can be extremely important in determining the nature of the reaction or the properties of the compound.

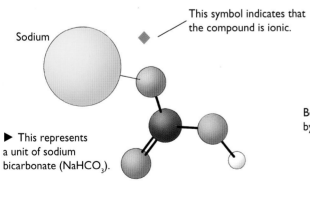

Sodium

This symbol indicates that the compound is ionic.

▶ This represents a unit of sodium bicarbonate ($NaHCO_3$).

The term "unit" is sometimes used to simplify the representation of a combination of ions.

Chemical symbols, equations and diagrams
The arrangement of any molecule or compound can be shown in one of the two ways shown below, depending on which gives the clearer picture. The left-hand diagram is called a ball-and-stick diagram because it uses rods and spheres to show the structure of the material. This example shows water, H_2O. There are two hydrogen atoms and one oxygen atom.

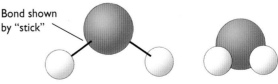

Bond shown by "stick"

Colours too
The colours of each of the particles help differentiate the elements involved. The diagram can then be matched to the written and symbolic equation given with the diagram. In the case above, oxygen is red and hydrogen is grey.

Glossary of technical terms

absorb: to soak up a substance. Compare to adsorb.

acetone: a petroleum-based solvent.

acid: compounds containing hydrogen which can attack and dissolve many substances. Acids are described as weak or strong, dilute or concentrated, mineral or organic.

acidity: a general term for the strength of an acid in a solution.

acid rain: rain that is contaminated by acid gases such as sulphur dioxide and nitrogen oxides released by pollution.

adsorb/adsorption: to "collect" gas molecules or other particles on to the *surface* of a substance. They are not chemically combined and can be removed. (The process is called "adsorption".) Compare to absorb.

alchemy: the traditional "art" of working with chemicals that prevailed through the Middle Ages. One of the main challenges of alchemy was to make gold from lead. Alchemy faded away as scientific chemistry was developed in the 17th century.

alkali: a base in solution.

alkaline: the opposite of acidic. Alkalis are bases that dissolve, and alkaline materials are called basic materials. Solutions of alkalis have a pH greater than 7.0 because they contain relatively few hydrogen ions.

alloy: a mixture of a metal and various other elements.

alpha particle: a stable combination of two protons and two neutrons, which is ejected from the nucleus of a radioactive atom as it decays. An alpha particle is also the nucleus of the atom of helium. If it captures two electrons it can become a neutral helium atom.

amalgam: a liquid alloy of mercury with another metal.

amino acid: amino acids are organic compounds that are the building blocks for the proteins in the body.

amorphous: a solid in which the atoms are not arranged regularly (i.e. "glassy"). Compare with crystalline.

amphoteric: a metal that will react with both acids and alkalis.

anhydrous: a substance from which water has been removed by heating. Many hydrated salts are crystalline. When they are heated and the water is driven off, the material changes to an anhydrous powder.

anion: a negatively charged atom or group of atoms.

anode: the negative terminal of a battery or the positive electrode of an electrolysis cell.

anodising: a process that uses the effect of electrolysis to make a surface corrosion-resistant.

antacid: a common name for any compound that reacts with stomach acid to neutralise it.

antioxidant: a substance that prevents oxidation of some other substance.

aqueous: a solid dissolved in water. Usually used as "aqueous solution".

atom: the smallest particle of an element.

atomic number: the number of electrons or the number of protons in an atom.

atomised: broken up into a very fine mist. The term is used in connection with sprays and engine fuel systems.

aurora: the "northern lights" and "southern lights" that show as coloured bands of light in the night sky at high latitudes. They are associated with the way cosmic rays interact with oxygen and nitrogen in the air.

basalt: an igneous rock with a low proportion of silica (usually below 55%). It has microscopically small crystals.

base: a compound that may be soapy to the touch and that can react with an acid in water to form a salt and water.

battery: a series of electrochemical cells.

bauxite: an ore of aluminium, of which about half is aluminium oxide.

becquerel: a unit of radiation equal to one nuclear disintegration per second.

beta particle: a form of radiation in which electrons are emitted from an atom as the nucleus breaks down.

bleach: a substance that removes stains from materials either by oxidising or reducing the staining compound.

boiling point: the temperature at which a liquid boils, changing from a liquid to a gas.

bond: chemical bonding is either a transfer or sharing of electrons by two or more atoms. There are a number of types of chemical bond, some very strong (such as covalent bonds), others weak (such as hydrogen bonds). Chemical bonds form because the linked molecule is more stable than the unlinked atoms from which it formed. For example, the hydrogen molecule (H_2) is more stable than single atoms of hydrogen, which is why hydrogen gas is always found as molecules of two hydrogen atoms.

brass: a metal alloy principally of copper and zinc.

brazing: a form of soldering, in which brass is used as the joining metal.

brine: a solution of salt (sodium chloride) in water.

bronze: an alloy principally of copper and tin.

buffer: a chemistry term meaning a mixture of substances in solution that resists a change in the acidity or alkalinity of the solution.

capillary action: the tendency of a liquid to be sucked into small spaces, such as between objects and through narrow-pore tubes. The force to do this comes from surface tension.

catalyst: a substance that speeds up a chemical reaction but itself remains unaltered at the end of the reaction.

cathode: the positive terminal of a battery or the negative electrode of an electrolysis cell.

cathodic protection: the technique of making the object that is to be protected from corrosion into the cathode of a cell. For example, a material, such as steel, is protected by coupling it with a more reactive metal, such as magnesium. Steel forms the cathode and magnesium the anode. Zinc protects steel in the same way.

cation: a positively charged atom or group of atoms.

caustic: a substance that can cause burns if it touches the skin.

cell: a vessel containing two electrodes and an electrolyte that can act as an electrical conductor.

ceramic: a material based on clay minerals, which has been heated so that it has chemically hardened.

chalk: a pure form of calcium carbonate made of the crushed bodies of microscopic sea creatures, such as plankton and algae.

change of state: a change between one of the three states of matter, solid, liquid and gas.

chlorination: adding chlorine to a substance.

cladding: a surface sheet of material designed to protect other materials from corrosion.

clay: a microscopically small plate-like mineral that makes up the bulk of many soils. It has a sticky feel when wet.

combustion: the special case of oxidisation of a substance where a considerable amount of heat and usually light are given out. Combustion is often referred to as "burning".

compound: a chemical consisting of two or more elements chemically bonded together. Calcium atoms can combine with carbon atoms and oxygen atoms to make calcium carbonate, a compound of all three atoms.

condensation nuclei: microscopic particles of dust, salt and other materials suspended in the air, which attract water molecules.

conduction: (i) the exchange of heat (heat conduction) by contact with another object or (ii) allowing the flow of electrons (electrical conduction).

convection: the exchange of heat energy with the surroundings produced by the flow of a fluid due to being heated or cooled.

corrosion: the *slow* decay of a substance resulting from contact with gases and liquids in the environment. The term is often applied to metals. Rust is the corrosion of iron.

corrosive: a substance, either an acid or an alkali, that *rapidly* attacks a wide range of other substances.

cosmic rays: particles that fly through space and bombard all atoms on the Earth's surface. When they interact with the atmosphere they produce showers of secondary particles.

covalent bond: the most common form of strong chemical bonding, which occurs when two atoms *share* electrons.

cracking: breaking down complex molecules into simpler components. It is a term particularly used in oil refining.

crude oil: a chemical mixture of petroleum liquids. Crude oil forms the raw material for an oil refinery.

crystal: a substance that has grown freely so that it can develop external faces. Compare with crystalline, where the atoms are not free to form individual crystals and amorphous where the atoms are arranged irregularly.

crystalline: the organisation of atoms into a rigid "honeycomb-like" pattern without distinct crystal faces.

crystal systems: seven patterns or systems into which all of the world's crystals can be grouped. They are: cubic, hexagonal, rhombohedral, tetragonal, orthorhombic, monoclinic and triclinic.

cubic crystal system: groupings of crystals that look like cubes.

curie: a unit of radiation. The amount of radiation emitted by 1 g of radium each second. (The curie is equal to 37 billion becquerels.)

current: an electric current is produced by a flow of electrons through a conducting solid or ions through a conducting liquid.

decay (radioactive decay): the way that a radioactive element changes into another element decause of loss of mass through radiation. For example uranium decays (changes) to lead.

decompose: to break down a substance (for example by heat or with the aid of a catalyst) into simpler components. In such a chemical reaction only one substance is involved.

dehydration: the removal of water from a substance by heating it, placing it in a dry atmosphere, or through the action of a drying agent.

density: the mass per unit volume (e.g. g/cc).

desertification: a process whereby a soil is allowed to become degraded to a state in which crops can no longer grow, i.e. desert-like. Chemical desertification is usually the result of contamination with halides because of poor irrigation practices.

detergent: a petroleum-based chemical that removes dirt.

diaphragm: a semipermeable membrane – a kind of ultra-fine mesh filter – that will allow only small ions to pass through. It is used in the electrolysis of brine.

diffusion: the slow mixing of one substance with another until the two substances are evenly mixed.

digestive tract: the system of the body that forms the pathway for food and its waste products. It begins at the mouth and includes the stomach and the intestines.

dilute acid: an acid whose concentration has been reduced by a large proportion of water.

diode: a semiconducting device that allows an electric current to flow in only one direction.

disinfectant: a chemical that kills bacteria and other microorganisms.

dissociate: to break apart. In the case of acids it means to break up forming hydrogen ions. This is an example of ionisation. Strong acids dissociate completely. Weak acids are not completely ionised and a solution of a weak acid has a relatively low concentration of hydrogen ions.

dissolve: to break down a substance in a solution without a resultant reaction.

distillation: the process of separating mixtures by condensing the vapours through cooling.

doping: adding metal atoms to a region of silicon to make it semiconducting.

dye: a coloured substance that will stick to another substance, so that both appear coloured.

electrode: a conductor that forms one terminal of a cell.

electrolysis: an electrical–chemical process that uses an electric current to cause the break up of a compound and the movement of metal ions in a solution. The process happens in many natural situations (as for example in rusting) and is also commonly used in industry for purifying (refining) metals or for plating metal objects with a fine, even metal coating.

electrolyte: a solution that conducts electricity.

electron: a tiny, negatively charged particle that is part of an atom. The flow of electrons through a solid material such as a wire produces an electric current.

electroplating: depositing a thin layer of a metal onto the surface of another substance using electrolysis.

element: a substance that cannot be decomposed into simpler substances by chemical means

emulsion: tiny droplets of one substance dispersed in another. A common oil in water emulsion is milk. The tiny droplets in an emulsion tend to come together, so another stabilising substance is often needed to wrap the particles of grease and oil in a stable coat. Soaps and detergents are such agents. Photographic film is an example of a solid emulsion.

endothermic reaction: a reaction that takes heat from the surroundings. The reaction of carbon monoxide with a metal oxide is an example.

enzyme: organic catalysts in the form of proteins in the body that speed up chemical reactions. Every living cell contains hundreds of enzymes, which ensure that the processes of life continue. Should enzymes be made inoperative, such as through mercury poisoning, then death follows.

ester: organic compounds, formed by the reaction of an alcohol with an acid, which often have a fruity taste.

evaporation: the change of state of a liquid to a gas. Evaporation happens below the boiling point and is used as a method of separating out the materials in a solution.

exothermic reaction: a reaction that gives heat to the surroundings. Many oxidation reactions, for example, give out heat.

explosive: a substance which, when a shock is applied to it, decomposes very rapidly, releasing a very large amount of heat and creating a large volume of gases as a shock wave.

extrusion: forming a shape by pushing it through a die. For example, toothpaste is extruded through the cap (die) of the toothpaste tube.

fallout: radioactive particles that reach the ground from radioactive materials in the atmosphere.

fat: semi-solid energy-rich compounds derived from plants or animals and which are made of carbon, hydrogen and oxygen. Scientists call these esters.

feldspar: a mineral consisting of sheets of aluminium silicate. This is the mineral from which the clay in soils is made.

fertile: able to provide the nutrients needed for unrestricted plant growth.

filtration: the separation of a liquid from a solid using a membrane with small holes.

fission: the breakdown of the structure of an atom, popularly called "splitting the atom" because the atom is split into approximately two other nuclei. This is different from, for example, the small change that happens when radioactivity is emitted.

fixation of nitrogen: the processes that natural organisms, such as bacteria, use to turn the nitrogen of the air into ammonium compounds.

fixing: making solid and liquid nitrogen-containing compounds from nitrogen gas. The compounds that are formed can be used as fertilisers.

fluid: able to flow; either a liquid or a gas.

fluorescent: a substance that gives out visible light when struck by invisible waves such as ultraviolet rays.

flux: a material used to make it easier for a liquid to flow. A flux dissolves metal oxides and so prevents a metal from oxidising while being heated.

foam: a substance that is sufficiently gelatinous to be able to contain bubbles of gas. The gas bulks up the substance, making it behave as though it were semi-rigid.

fossil fuels: hydrocarbon compounds that have been formed from buried plant and animal remains. High pressures and temperatures lasting over millions of years are required. The fossil fuels are coal, oil and natural gas.

fraction: a group of similar components of a mixture. In the petroleum industry the light fractions of crude oil are those with the smallest molecules, while the medium and heavy fractions have larger molecules.

free radical: a very reactive atom or group with a "spare" electron.

freezing point: the temperature at which a substance changes from a liquid to a solid. It is the same temperature as the melting point.

fuel: a concentrated form of chemical energy. The main sources of fuels (called fossil fuels because they were formed by geological processes) are coal, crude oil and natural gas. Products include methane, propane and gasoline. The fuel for stars and space vehicles is hydrogen.

fuel rods: rods of uranium or other radioactive material used as a fuel in nuclear power stations.

fuming: an unstable liquid that gives off a gas. Very concentrated acid solutions are often fuming solutions.

fungicide: any chemical that is designed to kill fungi and control the spread of fungal spores.

fusion: combining atoms to form a heavier atom.

galvanising: applying a thin zinc coating to protect another metal.

gamma rays: waves of radiation produced as the nucleus of a radioactive element rearranges itself into a tighter cluster of protons and neutrons. Gamma rays carry enough energy to damage living cells.

gangue: the unwanted material in an ore.

gas: a form of matter in which the molecules form no definite shape and are free to move about to fill any vessel they are put in.

gelatinous: a term meaning made with water. Because a gelatinous precipitate is mostly water, it is of a similar density to water and will float or lie suspended in the liquid.

gelling agent: a semi-solid jelly-like substance.

gemstone: a wide range of minerals valued by people, both as crystals (such as emerald) and as decorative stones (such as agate). There is no single chemical formula for a gemstone.

glass: a transparent silicate without any crystal growth. It has a glassy lustre and breaks with a curved fracture. Note that some minerals have all these features and are therefore natural glasses. Household glass is a synthetic silicate.

glucose: the most common of the natural sugars. It occurs as the polymer known as cellulose, the fibre in plants. Starch is also a form of glucose. The breakdown of glucose provides the energy that animals need for life.

granite: an igneous rock with a high proportion of silica (usually over 65%). It has well-developed large crystals. The largest pink, grey or white crystals are feldspar.

Greenhouse Effect: an increase of the global air temperature as a result of heat released from burning fossil fuels being absorbed by carbon dioxide in the atmosphere.

gypsum: the name for calcium sulphate. It is commonly found as Plaster of Paris and wallboards.

half-life: the time it takes for the radiation coming from a sample of a radioactive element to decrease by half.

halide: a salt of one of the halogens (fluorine, chlorine, bromine and iodine).

halite: the mineral made of sodium chloride.

halogen: one of a group of elements including chlorine, bromine, iodine and fluorine.

heat-producing: see exothermic reaction.

high explosive: a form of explosive that will only work when it receives a shock from another explosive. High explosives are much more powerful than ordinary explosives. Gunpowder is not a high explosive.

hydrate: a solid compound in crystalline form that contains molecular water. Hydrates commonly form when a solution of a soluble salt is evaporated. The water that forms part of a hydrate crystal is known as the "water of crystallization". It can usually be removed by heating, leaving an anhydrous salt.

hydration: the absorption of water by a substance. Hydrated materials are not "wet" but remain firm, apparently dry, solids. In some cases, hydration makes the substance change colour, in many other cases there is no colour change, simply a change in volume.

hydrocarbon: a compound in which only hydrogen and carbon atoms are present. Most fuels are hydrocarbons, as is the simple plastic polyethene (known as polythene).

hydrogen bond: a type of attractive force that holds one molecule to another. It is one of the weaker forms of intermolecular attractive force.

hydrothermal: a process in which hot water is involved. It is usually used in the context of rock formation because hot water and other fluids sent outwards from liquid magmas are important carriers of metals and the minerals that form gemstones.

igneous rock: a rock that has solidified from molten rock, either volcanic lava on the Earth's surface or magma deep underground. In either case the rock develops a network of interlocking crystals.

incendiary: a substance designed to cause burning.

indicator: a substance or mixture of substances that change colour with acidity or alkalinity.

inert: nonreactive.

infra-red radiation: a form of light radiation where the wavelength of the waves is slightly longer than visible light. Most heat radiation is in the infra-red band.

insoluble: a substance that will not dissolve.

ion: an atom, or group of atoms, that has gained or lost one or more electrons and so developed an electrical charge. Ions behave differently from electrically neutral atoms and molecules. They can move in an electric field, and they can also bind strongly to solvent molecules such as water. Positively charged ions are called cations; negatively charged ions are called anions. Ions carry electrical current through solutions.

ionic bond: the form of bonding that occurs between two ions when the ions have opposite charges. Sodium cations bond with chloride anions to form common salt (NaCl) when a salty solution is evaporated. Ionic bonds are strong bonds except in the presence of a solvent.

ionise: to break up neutral molecules into oppositely charged ions or to convert atoms into ions by the loss of electrons.

ionisation: a process that creates ions.

irrigation: the application of water to fields to help plants grow during times when natural rainfall is sparse.

isotope: atoms that have the same number of protons in their nucleus, but which have different masses; for example, carbon-12 and carbon-14.

latent heat: the amount of heat that is absorbed or released during the process of changing state between gas, liquid or solid. For example, heat is absorbed when a substance melts and it is released again when the substance solidifies.

latex: (the Latin word for "liquid") a suspension of small polymer particles in water. The rubber that flows from a rubber tree is a natural latex. Some synthetic polymers are made as latexes, allowing polymerisation to take place in water.

lava: the material that flows from a volcano.

limestone: a form of calcium carbonate rock that is often formed of lime mud. Most limestones are light grey and have abundant fossils.

liquid: a form of matter that has a fixed volume but no fixed shape.

lode: a deposit in which a number of veins of a metal found close together.

lustre: the shininess of a substance.

magma: the molten rock that forms a balloon-shaped chamber in the rock below a volcano. It is fed by rock moving upwards from below the crust.

marble: a form of limestone that has been "baked" while deep inside mountains. This has caused the limestone to melt and reform into small interlocking crystals, making marble harder than limestone.

mass: the amount of matter in an object. In everyday use, the word weight is often used to mean mass.

melting point: the temperature at which a substance changes state from a solid to a liquid. It is the same as freezing point.

membrane: a thin flexible sheet. A semipermeable membrane has microscopic holes of a size that will selectively allow some ions and molecules to pass through but hold others back. It thus acts as a kind of sieve.

meniscus: the curved surface of a liquid that forms when it rises in a small bore, or capillary tube. The meniscus is convex (bulges upwards) for mercury and is concave (sags downwards) for water.

metal: a substance with a lustre, the ability to conduct heat and electricity and which is not brittle.

metallic bonding: a kind of bonding in which atoms reside in a "sea" of mobile electrons. This type of bonding allows metals to be good conductors and means that they are not brittle

metamorphic rock: formed either from igneous or sedimentary rocks, by heat and or pressure. Metamorphic rocks form deep inside mountains during periods of mountain building. They result from the remelting of rocks during which process crystals are able to grow. Metamorphic rocks often show signs of banding and partial melting.

micronutrient: an element that the body requires in small amounts. Another term is trace element.

mineral: a solid substance made of just one element or chemical compound. Calcite is a mineral because it consists only of calcium carbonate, halite is a mineral because it contains only sodium chloride, quartz is a mineral because it consists of only silicon dioxide.

mineral acid: an acid that does not contain carbon and that attacks minerals. Hydrochloric, sulphuric and nitric acids are the main mineral acids.

mineral-laden: a solution close to saturation.

mixture: a material that can be separated out into two or more substances using physical means.

molecule: a group of two or more atoms held together by chemical bonds.

monoclinic system: a grouping of crystals that look like double-ended chisel blades.

monomer: a building block of a larger chain molecule ("mono" means one, "mer" means part).

mordant: any chemical that allows dyes to stick to other substances.

native metal: a pure form of a metal, not combined as a compound. Native metal is more common in poorly reactive elements than in those that are very reactive.

neutralisation: the reaction of acids and bases to produce a salt and water. The reaction causes hydrogen from the acid and hydroxide from the base to be changed to water. For

example, hydrochloric acid reacts with sodium hydroxide to form common salt and water. The term is more generally used for any reaction where the pH changes towards 7.0, which is the pH of a neutral solution.

neutron: a particle inside the nucleus of an atom that is neutral and has no charge.

noncombustible: a substance that will not burn.

noble metal: silver, gold, platinum, and mercury. These are the least reactive metals.

nuclear energy: the heat energy produced as part of the changes that take place in the core, or nucleus, of an element's atoms.

nuclear reactions: reactions that occur in the core, or nucleus of an atom.

nutrients: soluble ions that are essential to life.

octane: one of the substances contained in fuel.

ore: a rock containing enough of a useful substance to make mining it worthwhile.

organic acid: an acid containing carbon and hydrogen.

organic substance: a substance that contains carbon.

osmosis: a process where molecules of a liquid solvent move through a membrane (filter) from a region of low concentration to a region of high concentration of solute.

oxidation: a reaction in which the oxidising agent removes electrons. (Note that oxidising agents do not have to contain oxygen.)

oxide: a compound that includes oxygen and one other element.

oxidise: the process of gaining oxygen. This can be part of a controlled chemical reaction, or it can be the result of exposing a substance to the air, where oxidation (a form of corrosion) will occur slowly, perhaps over months or years.

oxidising agent: a substance that removes electrons from another substance (and therefore is itself reduced).

ozone: a form of oxygen whose molecules contain three atoms of oxygen. Ozone is regarded as a beneficial gas when high in the atmosphere because it blocks ultraviolet rays. It is a harmful gas when breathed in, so low level ozone, which is produced as part of city smog, is regarded as a form of pollution. The ozone layer is the uppermost part of the stratosphere.

pan: the name given to a shallow pond of liquid. Pans are mainly used for separating solutions by evaporation.

patina: a surface coating that develops on metals and protects them from further corrosion.

percolate: to move slowly through the pores of a rock.

period: a row in the Periodic Table.

Periodic Table: a chart organising elements by atomic number and chemical properties into groups and periods.

pesticide: any chemical that is designed to control pests (unwanted organisms) that are harmful to plants or animals.

petroleum: a natural mixture of a range of gases, liquids and solids derived from the decomposed remains of plants and animals.

pH: a measure of the hydrogen ion concentration in a liquid. Neutral is pH 7.0; numbers greater than this are alkaline, smaller numbers are acidic.

phosphor: any material that glows when energized by ultraviolet or electron beams such as in fluorescent tubes and cathode ray tubes. Phosphors, such as phosphorus, emit light after the source of excitation is cut off. This is why they glow in the dark. By contrast, fluorescors, such as fluorite, emit light only while they are being excited by ultraviolet light or an electron beam.

photon: a parcel of light energy.

photosynthesis: the process by which plants use the energy of the Sun to make the compounds they need for life. In photosynthesis, six molecules of carbon dioxide from the air combine with six molecules of water, forming one molecule of glucose (sugar) and releasing six molecules of oxygen back into the atmosphere.

pigment: any solid material used to give a liquid a colour.

placer deposit: a kind of ore body made of a sediment that contains fragments of gold ore eroded from a mother lode and transported by rivers and/or ocean currents.

plastic (material): a carbon-based material consisting of long chains (polymers) of simple molecules. The word plastic is commonly restricted to synthetic polymers.

plastic (property): a material is plastic if it can be made to change shape easily. Plastic materials will remain in the new shape. (Compare with elastic, a property where a material goes back to its original shape.)

plating: adding a thin coat of one material to another to make it resistant to corrosion.

playa: a dried-up lake bed that is covered with salt deposits. From the Spanish word for beach.

poison gas: a form of gas that is used intentionally to produce widespread injury and death. (Many gases are poisonous, which is why many chemical reactions are performed in laboratory fume chambers, but they are a byproduct of a reaction and not intended to cause harm.)

polymer: a compound that is made of long chains by combining molecules (called monomers) as repeating units. ("Poly" means many, "mer" means part).

polymerisation: a chemical reaction in which large numbers of similar molecules arrange themselves into large molecules, usually long chains. This process usually happens when there is a suitable catalyst present. For example, ethene reacts to form polythene in the presence of certain catalysts.

porous: a material containing many small holes or cracks. Quite often the pores are connected, and liquids, such as water or oil, can move through them.

precious metal: silver, gold, platinum, iridium, and palladium. Each is prized for its rarity. This category is the equivalent of precious stones, or gemstones, for minerals.

precipitate: tiny solid particles formed as a result of a chemical reaction between two liquids or gases.

preservative: a substance that prevents the natural organic decay processes from occurring. Many substances can be used safely for this purpose, including sulphites and nitrogen gas.

product: a substance produced by a chemical reaction.

protein: molecules that help to build tissue and bone and therefore make new body cells. Proteins contain amino acids.

proton: a positively charged particle in the nucleus of an atom that balances out the charge of the surrounding electrons

pyrite: "mineral of fire". This name comes from the fact that pyrite (iron sulphide) will give off sparks if struck with a stone.

pyrometallurgy: refining a metal from its ore using heat. A blast furnace or smelter is the main equipment used.

radiation: the exchange of energy with the surroundings through the transmission of waves or particles of energy. Radiation is a form of energy transfer that can happen through space; no intervening medium is required (as would be the case for conduction and convection).

radioactive: a material that emits radiation or particles from the nucleus of its atoms.

radioactive decay: a change in a radioactive element due to loss of mass through radiation. For example uranium decays (changes) to lead.

radioisotope: a shortened version of the phrase radioactive isotope.

radiotracer: a radioactive isotope that is added to a stable, nonradioactive material in order to trace how it moves and its concentration.

reaction: the recombination of two substances using parts of each substance to produce new substances.

reactivity: the tendency of a substance to react with other substances. The term is most widely used in comparing the reactivity of metals. Metals are arranged in a reactivity series.

reagent: a starting material for a reaction.

recycling: the reuse of a material to save the time and energy required to extract new material from the Earth and to conserve non-renewable resources.

redox reaction: a reaction that involves reduction and oxidation.

reducing agent: a substance that gives electrons to another substance. Carbon monoxide is a reducing agent when passed over copper oxide, turning it to copper and producing carbon dioxide gas. Similarly, iron oxide is reduced to iron in a blast furnace. Sulphur dioxide is a reducing agent, used for bleaching bread.

reduction: the removal of oxygen from a substance. See also: oxidation.

refining: separating a mixture into the simpler substances of which it is made. In the case of a rock, it means the extraction of the metal that is mixed up in the rock. In the case of oil it means separating out the fractions of which it is made.

refractive index: the property of a transparent material that controls the angle at which total internal reflection will occur. The greater the refractive index, the more reflective the material will be.

resin: natural or synthetic polymers that can be moulded into solid objects or spun into thread.

rust: the corrosion of iron and steel.

saline: a solution in which most of the dissolved matter is sodium chloride (common salt).

salinisation: the concentration of salts, especially sodium chloride, in the upper layers of a soil due to poor methods of irrigation.

salts: compounds, often involving a metal, that are the reaction products of acids and bases. (Note "salt" is also the common word for sodium chloride, common salt or table salt.)

saponification: the term for a reaction between a fat and a base that produces a soap.

saturated: a state where a liquid can hold no more of a substance. If any more of the substance is added, it will not dissolve.

saturated solution: a solution that holds the maximum possible amount of dissolved material. The amount of material in solution varies with the temperature; cold solutions can hold less dissolved solid material than hot solutions. Gases are more soluble in cold liquids than hot liquids.

sediment: material that settles out at the bottom of a liquid when it is still.

semiconductor: a material of intermediate conductivity. Semiconductor devices often use silicon when they are made as part of diodes, transistors or integrated circuits.

semipermeable membrane: a thin (membrane) of material that acts as a fine sieve, allowing small molecules to pass, but holding large molecules back.

silicate: a compound containing silicon and oxygen (known as silica).

sintering: a process that happens at moderately high temperatures in some compounds. Grains begin to fuse together even through they do not melt. The most widespread example of sintering happens during the firing of clays to make ceramics.

slag: a mixture of substances that are waste products of a furnace. Most slags are composed mainly of silicates.

smelting: roasting a substance in order to extract the metal contained in it.

smog: a mixture of smoke and fog. The term is used to describe city fogs in which there is a large proportion of particulate matter (tiny pieces of carbon from exhausts) and also a high concentration of sulphur and nitrogen gases and probably ozone.

soldering: joining together two pieces of metal using solder, an alloy with a low melting point.

solid: a form of matter where a substance has a definite shape.

soluble: a substance that will readily dissolve in a solvent.

solute: the substance that dissolves in a solution (e.g. sodium chloride in salt water).

solution: a mixture of a liquid and at least one other substance (e.g. salt water). Mixtures can be separated out by physical means, for example by evaporation and cooling.

solvent: the main substance in a solution (e.g. water in salt water).

spontaneous combustion: the effect of a very reactive material beginning to oxidise very quickly and bursting into flame.

stable: able to exist without changing into another substance.

stratosphere: the part of the Earth's atmosphere that lies immediately above the region in which clouds form. It occurs between 12 and 50 km above the Earth's surface.

strong acid: an acid that has completely dissociated (ionised) in water. Mineral acids are strong acids.

sublimation: the change of a substance from solid to gas, or vica versa, without going through a liquid phase.

substance: a type of material, including mixtures.

sulphate: a compound that includes sulphur and oxygen, for example, calcium sulphate or gypsum.

sulphide: a sulphur compound that contains no oxygen.

sulphite: a sulphur compound that contains less oxygen than a sulphate.

surface tension: the force that operates on the surface of a liquid, which makes it act as though it were covered with an invisible elastic film.

suspension: tiny particles suspended in a liquid.

synthetic: does not occur naturally, but has to be manufactured.

tarnish: a coating that develops as a result of the reaction between a metal and substances in the air. The most common form of tarnishing is a very thin transparent oxide coating.

thermonuclear reactions: reactions that occur within atoms due to fusion, releasing an immensely concentrated amount of energy.

thermoplastic: a plastic that will soften, can repeatedly be moulded it into shape on heating and will set into the moulded shape as it cools.

thermoset: a plastic that will set into a moulded shape as it cools, but which cannot be made soft by reheating.

titration: a process of dripping one liquid into another in order to find out the amount needed to cause a neutral solution. An indicator is used to signal change.

toxic: poisonous enough to cause death.

translucent: almost transparent.

transmutation: the change of one element into another.

vapour: the gaseous form of a substance that is normally a liquid. For example, water vapour is the gaseous form of liquid water.

vein: a mineral deposit different from, and usually cutting across, the surrounding rocks. Most mineral and metal-bearing veins are deposits filling fractures. The veins were filled by hot, mineral-rich waters rising upwards from liquid volcanic magma. They are important sources of many metals, such as silver and gold, and also minerals such as gemstones. Veins are usually narrow, and were best suited to hand-mining. They are less exploited in the modern machine age.

viscous: slow moving, syrupy. A liquid that has a low viscosity is said to be mobile.

vitreous: glass-like.

volatile: readily forms a gas.

vulcanisation: forming cross-links between polymer chains to increase the strength of the whole polymer. Rubbers are vulcanised using sulphur when making tyres and other strong materials.

weak acid: an acid that has only partly dissociated (ionised) in water. Most organic acids are weak acids.

weather: a term used by Earth scientists and derived from "weathering", meaning to react with water and gases of the environment.

weathering: the slow natural processes that break down rocks and reduce them to small fragments either by mechanical or chemical means.

welding: fusing two pieces of metal together using heat.

X-rays: a form of very short wave radiation.

Index